THE BRAW TREES OF COLDSTREAM

Also by Antony Chessell

The Life and Times of Abraham Hayward, Q.C., Victorian Essayist 'One of the two best read men in England' Lulu Publishing 2009

Editor, *A Small Share in the Conflict; The Wartime Diaries and Selected Correspondence of Flt Lt Henry Chessell (R.A.F. Intelligence Branch)* Lulu Publishing 2009

The Short and Simple Annals of the Poor; Some Family Ramblings Lulu Publishing, 2010 private circulation only

Coldstream Building Snippets   Cans, Quoins and Coursers Lulu Publishing 2010

THE BRAW TREES OF COLDSTREAM

Antony Chessell

Foreword by Lady Caroline Douglas-Home, MBE

Published by Lulu

First published in 2011

Lulu Enterprises UK Ltd., 7 Crooksbury Road

Farnham GU10 1QE www.lulu.com/uk

Copyright © Antony Chessell 2011

Antony Chessell asserts the moral right to be identified as the author of this work

A catalogue record for this book is available from the British Library

ISBN 978-1-4466-1735-9

Typeset in Times New Roman

Printed and bound in Great Britain

Front cover: Jacobs Well Wood, seen from Coldstream Bridge

Cover design and photograph by the author

To Athena Alice Robertson

whose namesake, the goddess of wisdom, created an olive tree to win the patronage of Athens

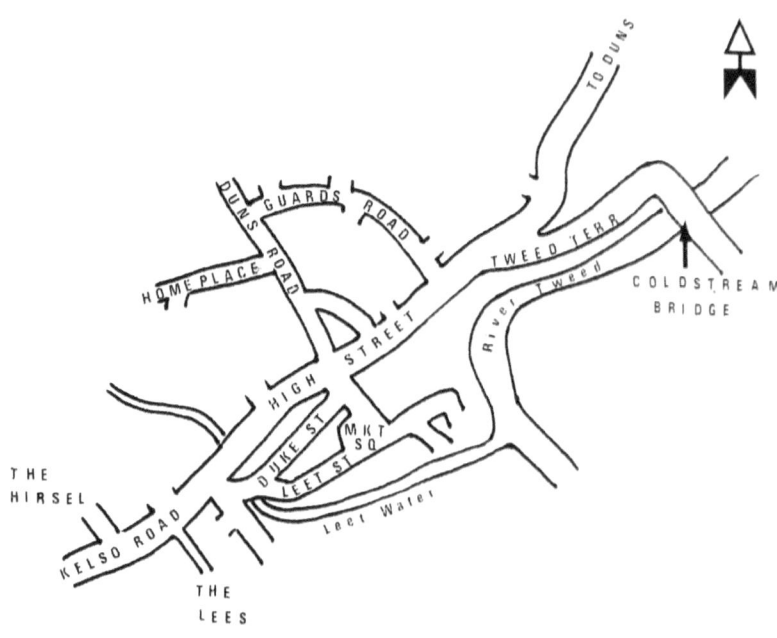

A simplified sketch map showing the older part of Coldstream

The distance covered by the map is about a mile from south-west to north-east. In the text, the locations of trees in the older part of the town can be found by reference to the street names on the map and from the photographs and accompanying descriptions. The locations of other trees on The Hirsel or The Lees estates, the entrances to which can also be seen on the map, are also described in the text and photographs. Eight figure Ordnance Survey grid references are given for all the trees.

CONTENTS

ILLUSTRATIONS

The photographs are too numerous to be listed in a separate table of illustrations and they just appear within the text:-

1. With unnumbered, explanatory captions, or

2. As 'thumbnails' with explanation in the surrounding text

All photographs are from the author's collection

The author stresses that he has not trespassed on any private property in order to take photographs. Most of the trees and areas of woodland are visible from roads, footpaths or other land with public access but in the few cases where it has been necessary to go on to private land, the owners have granted permission.*

*The Scottish Outdoor Access Code sets out rights and responsibilities for public access in terms of The Land Reform (Scotland) Act 2003, and lists the eight categories where there is no public right of access. One of these is houses and gardens, but if a house is located in a wider setting, Section 6 of the Act stipulates that there is no access on to land surrounding a house that is necessary to maintain privacy and undisturbed enjoyment of the house.

FOREWORD

By Lady Caroline Douglas-Home, MBE

In writing his excellent *Coldstream Snippets; Cans, Quoins and Coursers* Antony Chessell wrote of a subject of fascination to me. In *The Braw Trees of Coldstream* he has hit on an even longer interest. That the proceeds from both books are going to the Community Centre completes a trio of reasons why I am delighted to be writing this foreword.

My interest in trees was partly inherited but probably ignited mostly by living in a house in Birgham with a garden full of some definitely 'notable' trees. There was a very large Wellingtonia the bark of which meant it doubled as a punch-bag; a venerable beech which was a wonderful anchor for swings; several enormous limes with skirts which rabbits climbed into and made burrows through; old oaks and elms which housed nesting barn owls and many, many more.

I loved them all, whether with their bright spring growth meaning summer was round the corner, in full summer plumage showing what incredible varieties of green there are or as magnificent 'skeletons' against autumn and winter skies. Whatever the season trees really are beautiful and worth studying.

This book is a truly fascinating introduction to some of the trees in and around Coldstream, or wherever one is in the countryside. Antony has covered most aspects of them with professionalism, appreciation, feeling and readability. Follow him and discover them, and more, for yourself.

Caroline Douglas-Home   February 2011

INTRODUCTION

This is really a companion volume to my earlier *Coldstream Building Snippets: Cans, Quoins and Coursers* and, again, is for 'potterers' who walk around with their heads in the air, looking at things. Having written about the many building features that interested me in and around Coldstream, I wanted to move on to another local interest, trees, stimulated by my recording of 'ancient, veteran and notable' trees for the Woodland Trust's nationwide 'Ancient Tree Hunt' which aims to create a database of at least 100,000 trees by 2011—a sort of living version of listed buildings. I have always liked trees. As a small boy, I lived in Birmingham and was fortunate in going to nursery and primary school in Bourneville, which was a leafy 'garden village' created in the nineteenth century by the chocolate manufacturer and philanthropist, George Cadbury; trees were everywhere and formed a backdrop to my early life.

Later, in Croydon, my father's job meant that we lived in a rented flat in a Victorian house used for further education purposes, but which, before the war, had been the house occupied by the Principal of the Royal Normal College for the Blind. The former College grounds were laid out in lawns, embankments, flower-beds, trees and shrubs and, on one of the lawns, stood a mature *Catalpa bignonioides*, or Indian Bean tree ('Indian' from the once-named North American Indians). The tree seemed very exotic to my young mind and its large palmate leaves and long pendulous seed pods created an oasis of yellow-green dappled shade; it was also climbable and its branches were strong enough to take swinging ropes.

In Croydon in the 1940s and 1950s I joined the Wolf Cubs, Scouts and Senior Scouts and took part in many outdoor and camping activities in the English and Welsh countryside and mountains. Trees were part of an exciting landscape for adventures; they had practical uses for making shelters, camp kitchen and other gadgets from green wood and they provided the essential dead wood for firewood. At a mundane level, they even provided the sapling and leaf filters over the tops of grease pits for straining out the glutinous waste-water from cooking pots and soapy water from washing and washing-up bowls. We wore sheath knives and wielded axes but strict control ensured that no damage was done to mature trees and that felling and coppicing were confined to permitted areas and specific trees.

Many years later, my family and I lived in a converted croft in north-east Scotland and our own mixed woodland supplied fuel for our stove and a habitat for our free range hens and guinea fowl. Now we live in Coldstream where we can enjoy such a wide variety of trees in this unpolluted area; they are diverse, sturdy, spectacular and enduring—braw* trees in every way. Who could not be affected by them?

I wish to thank Lady Caroline Douglas-Home for graciously agreeing to write the foreword and for sharing her knowledge of Hirsel estate matters. My special thanks go to my wife, Gwen, for her help in recording the trees and for her constant encouragement.

Antony Chessell, Coldstream, Scottish Borders, 2011

*Scots for 'splendid, fine, upstanding, good' and similar complimentary epithets

Scots pine (*Pinus sylvestris*), Britain's only native pine; one of four to be seen to the north of Coldstream Primary School playing field. The Scots pine formed large parts of the original 'wildwood' in Britain and the ancient Caledonian forest in Scotland—see page 8. It can grow up to 36 metres tall with a girth of 1.50 metres and live for 150 years or more.

1

Setting the Scene

Having said that the influence of trees is strong in the Borders, aerial photographs, satellite images and hillside views of Coldstream and the eastern Borders might suggest otherwise. They show a rolling landscape of predominantly patchwork fields bounded by hedges with only pockets of trees lining rivers, roads and tracks or providing shelter-belts between fields or cover for game. It is true that

Looking north from the eastern flank of Branxton Hill looking towards The Hirsel woods and Coldstream.

there are areas of much greater tree cover on the large estates, due to managed planting over the centuries for ornamental as well as for sporting purposes, but the overall picture is of mainly arable land and pasture. Along the Tweed valley quite small hills or 'drumlins' formed by glacial deposits in the last Ice Age, add to the character of the rolling open landscape.

However, a closer look at ground level is more encouraging by revealing the extent of many of these tree pockets and shelter belts and the diversity of species within them; also the mixture of old or notable trees and more recently planted or regenerated trees. There is a rich variety within the categories of native and imported trees, whether hardwoods or softwoods, deciduous or evergreen, broadleaves or conifers. There is also encouraging evidence of good forestry and woodland management including new planting.

Hardwood trees have a higher density than softwoods and are slower growing—they can be deciduous (shedding their leaves in autumn) such as birch, sycamore, oak, beech and ash or they can be evergreen such as holly or Holm oak. Softwood trees have a lower

density, are faster growing and are mostly conifers such as pine, spruce, cedar, fir and larch. They are less expensive than hardwoods and are widely used in the building industry. Broadleaved trees are any deciduous or evergreen trees that have broad, flat leaves (see the beech leaves, above left), unlike conifers that are mostly evergreen (not larch) and have needle leaves (see the Stone pine needles, p.2 right) or scale leaves. So, the beech is deciduous, a hardwood and broadleaved; the Stone pine is evergreen, a softwood and has needles.

The *Supplementary Memorandum: EU Woodland* (House of Lords EU Sub-committee D) noted that in 2009 Scotland was one of the least wooded countries in Europe with tree cover equivalent to 17% of the total land area (England 9%, UK, 12%) compared with 37% in the European Union. Obvious natural reasons for this include climate, location and geography. Whereas the Scilly Isles and Lizard Point in Cornwall are south of the 50°N line of latitude, Scotland, including the northern isles, lies between 55°N and 62°N. Apart from having a colder climate than the rest of the United Kingdom (notwithstanding comparatively warm micro climates on, say, the west coast of Scotland or around the Moray Firth), the geography and altitude of mountain and moorland areas and windswept islands will also affect statistics relating to overall tree cover in Scotland.

But this is not the whole answer—for example, Forestry Commission Scotland in *Scotland's Trees, Woods and Forests*, 2002, lists Finland (between 60°N and 70°N) as having 72% tree cover and Sweden (between 55°N and 69°N) as having 66% cover. Also, thousands of years ago, Scotland had a much denser tree cover despite having these natural disadvantages. The main reason for the substantial decline in tree cover from 60% after the end of the last Ice Age has been human activity in terms of land clearance for farming, building construction and other domestic use of timber.

The *Scottish Borders Woodland Strategy*, 2005 (with *Moving Forward* Project 2.1, 2010 and *Further Developments*, 2009-2011), produced by Scottish Borders Council, gives more interesting national and regional figures relating to present-day tree cover. There are about 1.30 million hectares of woodland and forest in Scotland. Of these, 70% are conifer, 14% broadleaved and 4% mixed woodland with the balance being open space within woodland. Only about 2% of the total land in Scotland is semi-natural woodland so the balance of the 17% land cover represents trees that have been planted (or regenerated from planted trees) over the centuries. Today, the largest proportion of planted trees in Scotland comprises the commercial planting of conifers by private owners and the Forestry Commission (For. Com. Scotland since 2003) with a rapid expansion of planting after 1945. The dominant species is Sitka spruce which, in 2005, represented 48% of the total tree area; it grows quickly, provides a good quality building material and will be supplying at least 70% of softwood production for the next thirty years and beyond.

Woodland cover in the Scottish Borders is slightly higher than the above-mentioned national figure, being 18.5% of total land area and this has shown a dramatic increase over the 5% cover at the end of World War II. Tree cover in the Scottish Borders in 2005 was less than some parts of Scotland, e.g. Dumfries & Galloway was 26.8%, but was much more than Lothian which only had 10.5% cover; Scottish Borders had the fourth highest tree cover out of the nine regional council areas. The dominant species in the Borders is again Sitka spruce but it is important to realize that the figure for the Scottish Borders is skewed because it is the western Borders that has the large-scale commercial conifer plantations. To the east of the A7 and A6088, the eastern Borders has only small-scale privately owned plantations set in an environment of mainly agricultural land, much of it prime farming land. Tree cover is so much lower and it is this major

4

The edge of a Sitka spruce plantation. Usually having a dark and oppressive atmosphere, it is only the snow that is giving light to the scene here. But, Sitka spruce is a quick-growing and useful, commercial crop.

difference between the western and eastern Border areas that is reflected in the more open patchwork appearance of the landscape around Coldstream.

It was not always so. About ten thousand years ago at the end of the last Ice Age, according to Prof. Iain Stewart, ('Making Scotland's Landscape', BBC1, 2010), Scotland's 60% woodland cover was not adversely affected by its population of about 10,000 hunter-gatherers. These Mesolithic people trapped and killed wild animals and collected wild plants, nuts, fungi, and berries to provide enough food for themselves and their children in good times; in bad times they suffered starvation and death. They did not deplete woodlands and forests, only taking sufficient wood for shelter, simple tools and fuel, all the material for which was soon replaced by natural regeneration. The charity, Scotland's Native Woods, has put the maximum cover achieved by Scotland's ancient forest five thousand years ago at 80% or more but about four thousand years ago, according to Prof. Stewart, the climate became wetter and cooler and many trees drowned from oxygen starvation reducing the tree cover to about 25%. The growing of crops and other farming activities from Neolithic times onwards, resulted in an increase in population and a large-scale clearance of trees.

However, in addition to clearance, perhaps the management of trees also started in Neolithic times when farmers realised that hardwoods cut down to the ground for timber would sprout from the stump, producing many fast-growing thinner shoots used for everyday purposes and for items requiring smaller timber. The best results were obtained if cutting back or 'coppicing' was done every ten years or so on a rotational basis within the woodland and this is still best practice.

Standard trees were allowed to grow to maturity within the

coppice, providing a good mix of timber sizes. The higher light levels within a coppice also created a greater biodiversity of plants and animal life and allowed for regeneration from the seeds of mature trees. The tree shown here (see above) was coppiced a number of years ago and its damp position on the edge of The Hirsel lake has caused a vigorous multi-stemmed growth from the original stump. Unlike coppiced trees, 'pollarded' trees are cut at shoulder height so that the young shoots are

protected from grazing animals—the willow osier at The Hirsel lake (see above) is one of a number at the north-east corner of the lake that have been pollarded regularly for their useful cuttings.

There may have been a population of about 500,000 in Scotland in the Middle Ages and a tree cover of only 10% by the 16th century. By 1870, this had dropped still further to 4%, a level at which it stayed until only sixty years ago. The comparatively low level of tree cover in the UK including Scotland is therefore not a recent phenomenon but the decline was exacerbated during the 16th to 19th centuries by the surge in demand for timber for shipbuilding, particularly oak but also other woods that could not be easily met by existing supplies. For example, usable oak for the Royal Navy came from trees that had to be at least 200 years old. Even after the demise of timber hulls for warships and merchant ships, growing prosperity demanded more timber for commercial and domestic use and twentieth century wars boosted timber needs for all kinds of purposes including basic items such as thousands of pit props for the coal mining industry and poles and planks for the Flanders trenches.

The trees that are now considered to be native to Scotland arrived after the end of the last Ice Age when the ice sheet receded and disappeared over quite a short period of time—perhaps just a few decades. The warming bare soils were colonized by seeds from the south, the first pioneer plants being birch and willow followed by oak, elm, ash, hazel, alder, bird cherry, aspen, juniper, yew and Scots pine. These are all trees that are thought to have arrived without human intervention and would originally have formed what Professor Oliver Rackham has described as the original prehistoric forest or 'wildwood', a term originally coined in his *Ancient Woodland; its History, Vegetation and Uses in England*, Edward Arnold, 1980, and applied in his later works on British forests and woodlands and in the subsequent works of other authorities on the British countryside.

In Scotland, the 'wildwood' is usually identified with the ancient Caledonian Forest of Scots pine but the Forestry Commission in *Native Woodlands of Scotland*, 1998 lists other native trees that would have grown amongst them such as oak, ash, birch, rowan, aspen, gean (Wild cherry), hawthorn and juniper. The tree cover was not all impenetrable forest. It was probably thick in places and more open elsewhere with grassland areas between the trees according to ground and soil conditions. On higher ground or on the islands, the trees would have been stunted and limited to certain species such as willow and dwarf birch or small bent-over oak and hazel. Although the 'wildwood' no longer survives intact, traces of it may still exist in remote glens and on hillsides or in rocky gorges as 'semi-natural woodlands' or even 'ancient semi-natural woodlands'. They are not truly wild, having been affected to some extent by human activity but they are the nearest we have to the original forest landscape although, as mentioned on p. 4, 'semi-natural woodland' only forms 2% of Scotland's tree cover. In the Scottish Borders in 2005, this was even lower at only 0.26%.

The Forestry Commission has classified native woodland into six categories:-

1. Lowland and mixed broadleaved woods where oak predominates with ash and sometimes Wych elm with Silver birch and hazel growing in the underwood.

2. Upland mixed ashwoods usually confined to gullies and fertile damp ground.

3. Upland oakwoods characteristic of the western side of Scotland.

4. Upland birchwoods with different habitats in the north-west highlands or central and eastern highlands depending upon whether they are Silver birch or Downy birch.

5. Native pinewoods such as Scots pine in the Cairngorms.

6. Wet woodlands dominated by alder, willow or birch on wet or poorly drained soils.

Trees that did not colonize naturally are known as 'introduced species' having been brought in and planted here. Beech may or may not have been an introduced species in Scotland. It colonized its way slowly northwards from southern England and could have arrived in Scotland naturally but it may have received human help in doing so. Some introduced species were brought in by the great Scottish plant collectors in the 18th and 19th centuries, men such as David Douglas who collected more than 200 plant species in North America including the Douglas fir (*Pseudotsuga menziesii*). Exotic trees such as the Monkey Puzzle tree (the Chile pine, *Araucaria araucana*), the impressively tall Wellingtonia (*Sequoiadendron giganteum*) and many others were much in demand by owners of large country houses and estates who wanted them for ornamental landscape planting.

The most common introduced species for commercial timber production is the already mentioned Sitka spruce (*Picea sitchensis*) from West and North America. One introduced species, the sycamore (*Acer pseudoplatanus*) is so common and grows so well in Scotland that it is almost regarded as a native tree. There is no consensus on the date of its introduction; the Celts, the Romans and the Normans have been credited with bringing it from Europe (it should not be confused with other varieties such as the North American sycamore) and there have been other theories relating to the Middle Ages and to the reintroduction of the tree at different times in history. In Scotland, it can be known as the 'plane' tree; it looks rather like the London plane but *pseudo* fronts the species element of its name (see Note 2, p. 104).

In recent years, there has been a growing public awareness of the benefits of increasing tree cover, no more so than in Scotland. Government agencies and local authorities have developed policies and guidelines on woodland that are applied as mandatory requirements or as advisory good practice when decisions are taken in relation to a wide range of strategic and local matters such as Structure Plans, Local Plans and individual planning applications. Grants have also been made available in relation to new planting and regeneration projects. A number of charitable and supported organizations have arisen whose purpose is to protect existing woodland and encourage the purchase and planting up of new areas. These have included a good number of community groups where the input of local people has made a significant contribution not only in terms of awareness but also in the physical development of community woodland schemes. Some are mentioned in this book.

All of these organizations would, I am sure, support the benefits of tree cover as suggested by Scottish Borders Council in its *Scottish Borders Woodland Strategy*, 2005, p. 26:-

1. Improve air quality and climate.
2. Bind soil and reduce erosion.
3. Leaves contribute to soil quality.
4. Regulate run-off from surfaces.
5. Timber provides fibre, construction material and source of energy.
6. Environments for wildlife.
7. Scenic backdrop for living, recreation and health.
8. Screening and shelter.

These benefits form a background to the following chapters on some of the many fine trees to be seen in and around Coldstream.

2

Into Coldstream

There are four approach roads into Coldstream, all of them flanked by a verdant mixture of farmland, hedges and trees. From the west, the arable, hedged landscape gives way to  overhanging trees of The Hirsel and The Lees estates on either side of Kelso Road in the vicinity of the Health Centre (see left). From the north, the famous Duns Road 'switchback' passes farmland and the wooded Hirsel estate and becomes tree-lined within the hedge-line or in gardens between the boundary of the built-up area and Guards Road. From the east, the Lennel Road is quite heavily wooded until it meets the built-up area when it then curves down to meet the High Street.

The most dramatic approach to the town is from the south, crossing over John Smeaton's old Coldstream Bridge, built between 1763 and 1766, which (on foot) gives such dramatic views of the sweeping bends of the River Tweed, upstream and of the foaming but protective (from foundation scouring) *cauld* or weir, downstream. On leaving England, half way across the bridge, the first sight of Scotland is of a wooded area owned by the Woodland Trust on the north bank of the River Tweed. As the road swings round before entering Coldstream, this steeply sloping area is immediately on the left hand side. The Gallowsknowe and Belmont woodlands are across the road.

Woodland Trust land, on the north side of the River Tweed. Contrast this winter scene with the summer scene shown on the front cover.

The Woodland Trust land of one hectare (2.47 acres), known as Jacob's Well Wood, used to belong to the Lennel estate but was gifted to the Trust in 1995. The Trust is a charity that wants to see more native woods and trees in Britain. The Ordnance Survey map of 1858 shows only a few trees here and aerial photographs taken by the RAF between 1944 and 1950 (Royal Commission on the Ancient and Historical Monuments of Scotland) do not show this as being wooded, apart from trees along the line of the road and some around Jacob's Well. The Trust's records show that in 1989, 0.50 hectare was felled and restocked with mixed broadleaved trees at a density of 1100 trees per hectare which is a fairly low planting density with trees approximately 3.50 metres apart.

The land is now well wooded, the western part consisting mainly of sycamore on very steep ground above the River Tweed and the eastern part consisting of a closed canopy (where the crowns of trees overlap to form a continuous layer) of mainly ash that was thinned in 2004. The trees, particularly those on the steep slopes above the river, help to bind the soil and prevent landslides. It is a fairly narrow strip of land varying in width between 20 metres and 40 metres. The wood is part of the River Tweed Site of Special Scientific Interest (SSSI) designated in 1976 and re-designated in 2001. The SSSI became a Species Area of Conservation (SAC) in 2005. This is because of the landscape value of the riverside woodland along the whole of the Tweed valley and the quality of the freshwater habitat for lamprey, otter and salmon.

Jacob's Well itself is served by a natural spring channelled into a pipe which discharges into a rough, circular, stone-sided shallow depression that, at one time, may have had greater depth.

14

Steep steps lead down from a gap in the wall beside the main road to a path that crosses over a short boardwalk above the well.

The steps and path down to the boardwalk with handrail that crosses over Jacob's Well; a covering of autumn leaves obscures the lower steps.

The overflow from the well runs out into an unlined channel down to the river. The origin of the name, Jacob's Well, is unknown but the Trust believes that there may have been a biblical connection taken from the story of Christ meeting a Samaritan woman at the well that Jacob gave to his son, Joshua (John 4:6).

Jacob's Well, partly hidden under the boardwalk.

In times gone by, washerwomen used to gather at the well but the spring water was also good and plentiful enough to be collected and used for whisky blending by R. Carmichael & Sons in the High Street. Coldstream Brewery (Messrs. J. A. Davidson, later Richard Day), owners of the Newcastle Arms, the Red Lion and the grocery shop in Duke Street, also blended whisky but the brewery probably had water on site, without having to use Jacob's Well.

Moving on into the town, there are not many trees alongside roads and streets as this was not traditional in the market squares or thoroughfares of Scottish towns and villages. Even in the unusually large Market Square in Kelso, the cobbles are uninterrupted by any tree planting. In Coldstream, there are exceptions where, for example,

Scottish Borders Council and their predecessors have carried out street planting in conjunction with municipal housing schemes such as the ornamental trees at Bookie Lane (see left) and those near the junction of Guards Road and Priory Hill. Also, a few years ago, the Council planted some Small-leaved lime trees in Market Square interspersed by raised

flowerbeds (see right). This has had a softening effect on the paved central area. Small-leaved lime (*Tilia cordata*) has, as its name suggests, leaves that are smaller than its relatives, the Large-leaved lime (*Tilia platyphyllos*) and the Common lime (*Tilia x europaea*). It is a deciduous broadleaf and its flowers, like the other limes, are

sweet-smelling and attract so many insects that the trees are said to hum or buzz in the summer. In the wild, Small-leaved limes were coppiced for their tight-grained wood that was used for fuel and for

17

making spoons, ladles and other domestic items. In some parts of Britain, its leaves were once used as cattle fodder and the blossom was used as a substitute for tea during the Second World War. It was obviously a very useful tree and the fibrous under bark was even twisted into rope or made into sandals.

Coldstream is fortunate in that, although it does not have avenues of trees along its streets (and there can be downsides because

of root damage and maintenance costs), there are many green areas such as the playing fields and Henderson Park; also substantial mature gardens with trees at the rear of listed properties within the central conservation area. There is the tree-lined River Leet (see left with the mature weeping willow) winding down from The Hirsel estate under the High Street and the Irish Bridge past the ends of Duke Street and Leet Street to its confluence with the River Tweed at the historic Tweed Green with its river frontage. The true weeping willow (*Salix babylonica*) comes from China and is rare in Europe but hybrids with White willow (*Salix alba*) or Crack willow (*Salix pendulina*) are common in Britain. The hybrid forms are *Salix sepulcralis* and *Salix pendulina* respectively; the first of these is the most common and has yellow drooping shoots.

So, Coldstream has many green 'lungs' and trees can be seen throughout the town, many of them in private gardens; some of the larger ones are survivors of a time when the land on which they stand was part of a larger estate or landholding. For example, between the

wars, Coldstream Burgh Council built a row of semi-detached houses at Home Terrace in Duns Road. This would have been on land that was originally part of The Hirsel estate, the seat of

the Earls of Home. The mature oak trees lining the side of Duns Road would have been on the line of a former field boundary of The Hirsel

at a time when Duns Road was not pedestrianised at Toun Heid. The trees would have created a leafy approach to the town as well as a pleasant frontage between the road and the gardens in front of the new development. These trees have now gone although two

stumps can still be seen set into front hedges (see above). Further up Duns Road between the junction with Guards Road and Bennecourt Drive, the line of old oaks still survives along or behind the front hedges of the private houses (see below). There will be other examples of many old trees that survive in the town, not necessarily on road or street frontages, but some that were planted centuries ago

in the private gardens of old houses within the conservation area or that were already there when existing plots were divided or when new estates were developed around them.

Before the expansion of Coldstream from the 19th century onwards, there were many more trees in the *toun* than there are now. Houses, shops, kirk, manse, brewery and municipal buildings were all crowded into a small area around Market Place, the roads leading off it, the bottom part of the road leading to Duns and the western and central parts of what is now the High Street. The estates of the Earl of Home, the Marjoribanks family and the Earl of Haddington not only encircled the central areas but included much of the village itself. Sharp's map of Coldstream in 1818 shows this quite clearly but it also shows the amount of garden ground behind the buildings, the large areas used as fruit gardens and the many trees throughout the village.

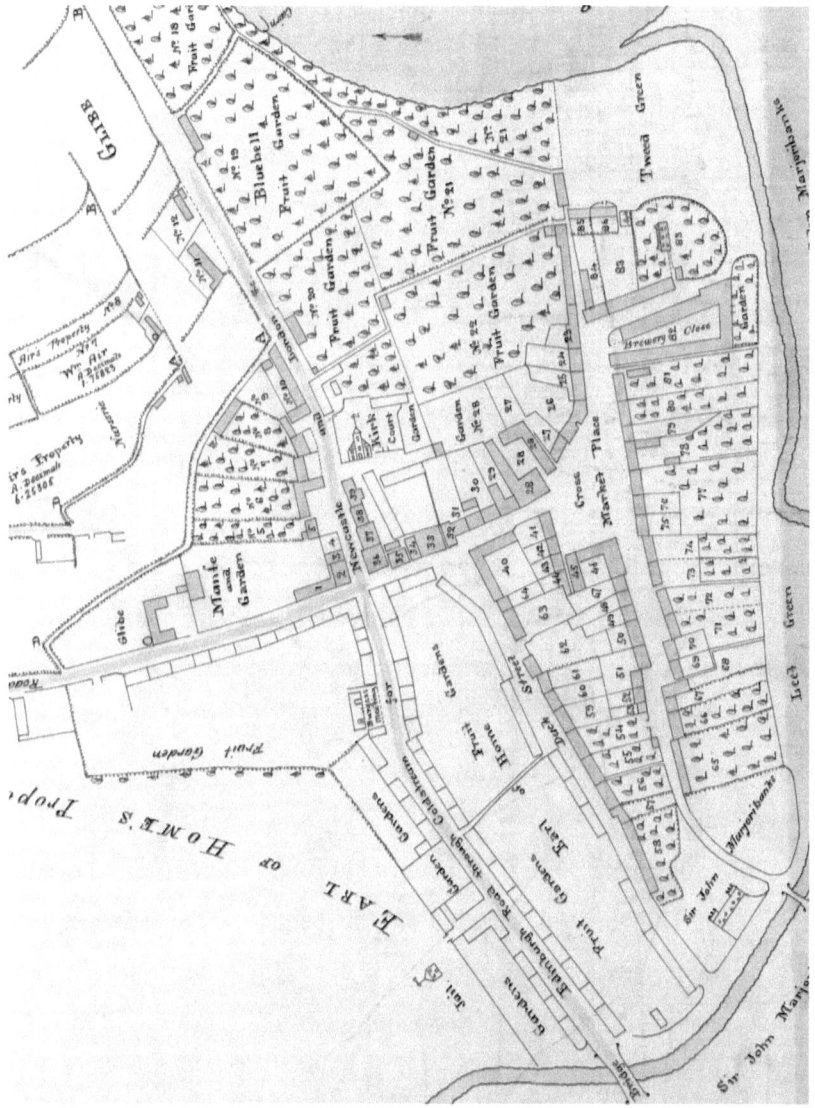

Extract (rotated, with north ←) from Peter Sharp's map of Coldstream, circa 1818, (NLS shelfmark EMS.s.372) showing the many gardens, fruit gardens and orchards. Fruit gardens 20 and 21 can be seen immediately to the west (↓) of the Bluebell Garden.   (Reproduced by kind permission of the Trustees of the National Library of Scotland).

In 2007, the Borders Forest Trust, a charity established in 1996 to conserve, restore and manage native woodland including local orchards, commissioned a *Survey of Border Orchards* by botanist, Louise Seed. In Coldstream, it was discovered that only traces of the orchards survive at Tweed Villa, Bank House and Orchard Cottage but it is clear that, historically, Coldstream was an important fruit-growing village because it was ideally situated with gentle south sloping gardens bounded by the River Tweed and the River Leet. The orchard at Tweed Villa formerly belonged to Rosybank House before the latter was partly converted into flats and, according to research carried out by Mr. James Hardy for the Berwickshire Naturalists Society in 1876, was on the site of the *pomarium* (medieval term for an apple orchard) of Coldstream Priory.

The survey gives a list of old varieties of pear and apple that were there when the orchard was part of Rosybank House. The orchard is shown on the 1818 map as No. 21. The garden may therefore have been used as an orchard for over 800 years. The 2007 survey comments:

> Coldstream priory, founded in 1150, by Gospatrick, the Earl of Dunbar and situated on the banks of the Tweed, was important for fruit growing. In the 13th century, in a return for papal taxation in Scotland, Coldstream priory is the only ecclesiastical establishment that derived revenue from the produce of its fruit gardens (Hardy, James, *Notice of the Orchard of Coldstream Priory* Transactions of the Berwickshire Naturalists' Club, 1876-1878, pp. 312/313). In the Borders, as in the rest of Scotland, monasteries and private estates were the main practitioners of growing fruit

trees until the 20th century (Butterworth, John, *Apples in Scotland,* Longford Press, 2001).

The orchard at Bank House is shown on the 1818 map as No. 20.

In the following chapters, I am going to highlight some notable, veteran and ancient trees that appeal to me; it will be a selective and personal choice although I will try to explain my reasons. It will not be a gazetteer or a botanical index of trees in Coldstream—there are far too many and this is not a textbook. I will describe individual specimens and mention the characteristics of different species but this can only give a flavour of the richness and variety of trees to be seen in this area.

3

Some Notable Trees

The Woodland Trust has come up with three categories for trees that might be regarded as special because of their age, their history, their location, or their intrinsic characteristics (www.ancient-tree-hunt.org.uk/ancienttrees):-

1. A 'notable tree' is one that is 'a tree of local importance, or of personal significance to the individual recorder. This includes specimen trees or those considered to be potential, next generation, veteran trees'.

2. A 'veteran tree' is one that is 'usually in the second or mature stage of its life and has important wildlife and habitat features including hollowing or associated decay, fungi, holes, wounds and large dead branches. It will generally include old trees but also younger, middle aged trees where premature ageing characteristics are present'.

3. An 'ancient tree' is one that is 'of interest biologically, aesthetically or culturally because of its age'. It will be 'in the ancient or third or final stage of its life'. It will be 'old relative to others of the same species'. The Trust explains that 'an ancient tree is one that makes you go "wow, it's huge, fatter than any tree like that round here". It will be really fat, but probably not that tall, as like old people they shrink down with age. Like people, trees grow and age at different

rates depending on where they are and what happens to them during their lifetime. Ancient trees are living relics of incredible age that inspire in us feelings of awe and mystery. They also support wildlife that cannot live anywhere else. Over the centuries, they have inspired artists, writers, poets and scientists and are mentioned in sacred texts'.

In this chapter, I have selected a number of 'notable trees' that stand out for me. Bearing in mind the many thousands of trees in and around Coldstream, it can only be a subjective choice and should only be regarded as a representative selection—there are hundreds more that appeal to me and there will be many more that appeal to other people for a variety of reasons that are personal to them.

My first tree is a solitary lime tree on Tweed Green (Grid Ref. NT84433971). The OS map of 1858 (National Library of Scotland) shows a line of trees along a boundary on Tweed Green. A late Victorian photograph in John Griffiths' book *Old Coldstream and Cornhill* shows three trees along a fence. The Canmore Collection (Royal Commission on the Ancient Monuments of Scotland) shows three or four trees after 1944 (Item SC 1020085) and two trees in 1999 (Item SC 1018785). However, the lime was probably not there in 1818 because Sharp's map (National Library of Scotland) shows Tweed Green, north of Tweed Road, as one of the many fruit gardens in Coldstream, including the neighbouring Bluebell fruit garden.

This solitary tree is therefore a Victorian survivor and one to be admired considering its exposed position and the fact that it has to stand in several feet of water whenever the River Tweed floods and covers Tweed Green. Another reason why I have included it is because of the historic site on which it stands. Tweed Green was part of, or next to, Coldstream Priory and, after the battle of Flodden Field

The solitary lime tree on Tweed Green.

in 1513, the Abbess Hoppringle sent for the bodies of the Scottish nobility so that they could be brought back for burial in consecrated ground at the priory. A pen and ink map of 1589 (The National Archives of Scotland ID RHP 49993) shows the priory kirkyard on land between the River Tweed and the mouth of the River Leet. The tree may be a little to the north of the site of the kirkyard but it is impossible to be precise as the map is not to an exact scale. Tweed Green has never been the subject of archaeological investigation but,

26

if it is a burial site, at least it has been left undisturbed. Every August, towards the close of Flodden Day, the area next to the tree is the site of a moving and poignant ceremony when the Coldstreamer and his attendants bring back the sod of earth from the battlefield on Branxton Hill. The land is also not far from the ancient ford over the river used by Scottish and English armies on their forays north and south.

The tree is probably a Common lime (*Tilia x europaea*) rather than a Large-leaved lime (*Tilia platyphyllos*), but it is one or the other. The Common lime is a hybrid of the Large-leaved lime and the Small-leaved lime and has a natural propensity to produce what is known as *epicormic* growth, which is the thick twiggy growth around the base of the trunk. Because of this almost impenetrable growth, I certainly had difficulty in measuring the girth of the tree (2.59 metres, measured at 1.50 metres above ground level). This is not an exceptional size for the species and is not the reason for choosing it.

Both the Large-leaved lime and the Small-leaved lime are native to Britain (but not Scotland) but are only two of over thirty species in the northern hemisphere. Like the Small-leaved lime, the Large-leaved lime wood has a tight grain and is easily workable, being used for intricate carving, for example for guitar manufacture and model making. Lime trees are very valuable to beekeepers but motorists would be wise not to park underneath one in the summer because of the sticky 'honeydew' dropped from the leaves by aphids.

There are many oak trees in and around Coldstream and probably most of these are Pedunculate or English oaks (*Quercus robur*), rather than Sessile oaks (*Quercus petraea*), that predominate on higher ground and on poorer and drier soils. However, the planting of trees rather than natural regeneration can blur the distribution of all

species. Oaks will feature in later chapters dealing with veteran and
ancient trees but here I wanted to feature three interesting oaks that do

Oak tree 'sentinel' at the entrance to The Hirsel estate, Kelso Road.

not fall into these categories but appeal to me on other grounds. The first two deserve some recognition simply because of the 'sentinel' positions that they occupy. The first is at the entrance to The Hirsel estate on the grassed area between the lodge and the Kelso Road (Grid Ref. NT83753943). It has a fairly large girth (4.00 metres at 1.50 metres above ground) but not many large upper branches; these have suffered by being so close to the main road. This view in winter (see p. 28) shows that it has been heavily pruned in the past to remove overhanging branches. It looks rather 'skeletal' without its foliage.

Oak tree 'sentinel' at the entrance to Coldstream Health Centre.

The second oak occupies a similar position on a grassed area at the entrance to Coldstream Health Centre, Kelso Road (Grid Ref. NT83493935). It is not as wide (girth of 3.62 metres at 1.00 metre height) but it has a chunky and solid appearance with a 'twisted' look to its upper branches. Both these trees are probably not noticed much by sick or healthy pedestrians and passing motorists (even at a speed below 30 mph), whose minds are on other things.

'Coronation Oak' planted in 1902 to commemorate the Coronation of King Edward VII.

The third oak is the 'Coronation Oak' on The Hirsel estate outside the east wall of the walled garden (Grid Ref. NT82924053). It was planted in 1902 during the time of Charles Alexander Douglas Home, 12th Earl of Home, great grandfather of the present Earl. The

tree has a girth of 2.90 metres at a height of 1.50 metres above ground level and, as the tree was planted as an acorn in 1902, it shows the extent of the growth achievable by an oak in 109 years. As mentioned (p. 8), the minimum age for an oak tree before it was suitable for building ships of the line for the Royal Navy was 200 years so this tree is only just over half way there; fortunately for the tree, it was planted 300 years too late.

Ditton Park, Buckinghamshire, where the acorn came from in 1902, was a seat of the Montagu family and the connection with the Earls of Home goes back to the marriage in 1798 of the 10th Earl of Home and Lady Elizabeth Scott, who was the daughter of the 3rd Duke of Buccleuch and Lady Elizabeth Montagu, the daughter of the 1st Duke of Montagu. The connection was strengthened when, in 1832, the 11th Earl of Home married his cousin, the Hon. Lucy Elizabeth Montagu, daughter of the 2nd Baron Montagu and the Hon. Jane Margaret Douglas. The 11th Earl was created Baron Douglas of Douglas in the County of Lanark because the Hon. Jane Douglas was the daughter of the 1st Baron Douglas. Her son, the 12th Earl of Home, took the additional name of Douglas on the death of his mother in 1877.

Pedunculate and Sessile oaks are native, deciduous, broad-leaved hardwoods. There are two other species of oak in Britain that are introduced species. These are, firstly, Turkey oak (*Quercus cerris*), a deciduous, longer-leaved variety from south-east France, the Balkans and across into Turkey and, secondly, Holm, or Holly, oak (*Quercus ilex*) from the Mediterranean area, an evergreen variety introduced into Britain, perhaps in the 1500s. I mention them even though they are seen more in southern areas of Britain.

Perhaps the easiest way to distinguish between Pedunculate oak and Sessile oak is to note that the former has short-stalked leaves and long-stalked acorns, whereas the latter has long-stalked leaves and short-stalked acorns. However, just to confuse matters, hybrid trees with intermediate characteristics can occur where both species are in close proximity. Oak is very hard, strong, decorative and extremely durable outdoors and does not need any preservative treatment due to the presence of *tannins* in the heartwood that make the timber extremely resistant to decay. For these reasons, oak was always an important building material as well as the main shipbuilding timber and it can last for hundreds of years without painting or maintenance. It turns grey on exposure to the atmosphere and increases in strength with the passage of time.

In addition, oak has always been used for making high quality durable furniture and its bark was also used for tanning leather because the tannins protect hides from decay; its acorns were traditionally used for fattening up pigs, being a very rich food source.

Wellingtonia at The Lees estate, close to Lees Lodge.

My selection of notable trees in Coldstream does include some introduced species, particularly as they have adapted well to Coldstream soils and climatic conditions. One prominent tree is the

33

magnificent Wellingtonia (*Sequoiadendron giganteum*) at The Lees. In its natural habitat in California, the tree can grow to 85 metres and even in Britain it can grow to 50 metres or more and growth is very rapid because this height is achieved after only 150 years. Its introduction was by a Scotsman, John D. Matthew, who collected a small quantity of seeds in the Calaveras Grove, California, bringing them to Scotland in August 1853 for distribution amongst friends.

Shortly afterwards, another plant hunter, William Lobb, was employed by Veitch Nurseries of Exeter and it was while he was on a collecting trip in California that he recognized the commercial prospects of bringing Wellingtonia seeds to Britain for ornamental planting. So did his employer, James Veitch, who passed samples to John Lindley, Professor of Botany at the University of London, for classification. The name given by Lindley, *Wellingtonia gigantea*, was not valid because of the already existing *Wellingtonia arnottiana* and it was later renamed *Sequoiadendron giganteum*. However, the name Wellingtonia remained as the popular name in Britain having originally been chosen in memory of the victor of Waterloo, the 1st Duke of Wellington who had recently died in 1852.

The tree shown on page 33 is a particularly fine specimen just off the drive up to The Lees and close to Lees Lodge on the Kelso Road (Grid Ref. NT83853943). It has a girth of 8.12 metres at a height of 1.50 metres above ground and is the widest Wellingtonia that I have found in and around Coldstream; it is not the widest in Scotland as this is claimed by Cluny House Gardens in Perthshire which has a 130 years old specimen with a girth of 11.00 metres. There are many examples of Wellingtonia on The Hirsel and The Lees estates and there are isolated trees elsewhere, such as the one in the garden of the Old Manse at Home Place. At The Hirsel, they were often planted in avenues, the most notable being along the old access to the house from the Castlelaw side. There are also some alongside the road in to The Hirsel Golf Club. Note the tall, very straight trunk.

The Royal Botanic Gardens at Kew (www.kew.org/plants-fungi/Sequoiadendron-giganteum.htm) places the Wellingtonia as the world's largest tree by volume and able to live for up to 3,200 years in its native habitat on the western slopes of the Sierra Nevada in California. However large and impressive the specimens in Coldstream, they are still young trees and can therefore only be included in the 'notable' category—they are not 'ancient', or even 'veteran'. The botanical name *Sequoia* comes from the Cherokee Indian, Sequoyah, who invented the Cherokee alphabet, adopted by the tribe in 1825; *dendron* is Greek for tree. The Wellingtonia is also known as the Giant Redwood because of the colour of its bark, which is fibrous and spongy to the touch. Because the bark can be over two feet thick, the tree is able to withstand fire damage and fire is actually an aid to opening the seed pods and scattering the seeds. The wood has no commercial use and the value of the tree is in its ornamental use that appealed so much to Victorian landowners. Wellingtonia trees can be seen in estate and parkland settings all over Britain, with many fine examples across the Scottish Borders.

Coldstream is fortunate in having a wide variety of ornamental trees planted in the nineteenth century by major landowners. Many of the plant hunters were Scottish and they brought back what were considered to be exotic trees, shrubs and other plants from all over the world. Conifers of all kinds were much in demand and, 150 years or so later, they have grown towards maturity and reached heights and girths that could only have been imagined by their planters. They include species such as the Douglas fir, Grand fir, Noble fir, Silver fir, Eastern hemlock, Western hemlock, Lawson's cypress, Nootka cypress, Atlas cedar, Cedar of Lebanon, Western red cedar, Lodgepole pine and Japanese larch. There are many others but I mention the above species because all of them are represented at The Hirsel. These and others that I have not listed, are notable for the way they have adapted and become successful in the woodland landscapes across Scotland.

Three of them can be seen quite easily in identifiable locations at The Hirsel without having to tramp through woodland on a frustrating search. The Silver fir (*Abies alba*) is a beautiful, tall tree and is a native of mountainous areas in central and southern Europe from the Pyrenees in the west to the Alps, the Carpathians and down into Serbia. It has a straight trunk and grows up to 50 or 60 metres; it was used as the main Christmas tree before the Norway spruce became more popular. The crown of a mature tree has a blunt or flattened top known as a 'stork nest crown'. The needles are dark green above and have two narrow white lines underneath that are lines of stomata or pores. It has a fairly soft whitewood that is used for building and carpentry and its oil is used in aromatherapy treatments. The tree has very deep roots enabling it to have a very firm anchor in poor mountainous soils; very few trees have deeper root systems. It is a member of the pine family of conifers.

The tree shown here is next to the path that starts opposite the Homestead Visitor Centre and runs along the south side of the lake (Grid Ref. NT82624016). It is one of three trees at this spot—one of them is dead and all have suffered wind damage, including this one.

36

Silver fir near The Hirsel lake, showing wind damaged branches.

It has a girth of 3.49 metres at a height above ground of 1.50 metres. There is a thick understorey here in the damp conditions near the lake with a hazel 'hedge' and many other species of trees and shrubs, some native and others not, such as the invasive rhododendron.

Atlas cedar next to the drive and near the main entrance to The Hirsel.

The Atlas cedar (*Cedrus atlantica*) (see preceding page), a member of the pine family, is another mountain species which, as its name suggests, comes from the Atlas mountains of Algeria where it

 grows up to 50 metres—but only about 30 metres in Europe. It is a straight tree with branches angled upwards and with an irregular top. Its blue-green needles are single on long shoots and in whorls on short side shoots and its cones have a dense barrel shaped appearance (see above). It is a very impressive tree and, although it can be used for building and furniture making, it is mostly planted for ornamental use in situations that provide plenty of growing space. Like other cedars, its oil is used for aromatherapy. The tree shown on the preceding page is quite large with a girth of 5.30 metres at a height of 1.50 metres above ground level. It is easy to find, just beyond the statue of Sir Alec Douglas-Home (14th Earl of Home) at Grid Ref. NT83673950.

Another cedar, the Western Red cedar (*Thuga plicata*), is not from the same family as the Atlas cedar, as it is a type of cypress rather than a pine, although they are closely related. Beyond a difference in DNA, there are subtle recognizable differences, one of these being that most cypress species have flattened, scale-like leaves, sometimes very short, whereas pines tend to have longer leaves that are round or polygonal in cross-section. The one I have singled out can be seen outside the entrance to The Hirsel estate office at the north-west corner of the walled garden (Grid Ref. NT82814053). The tree originates from the Pacific coast of North America and is another large tree growing up to 60 metres in its native habitat but only about

Western Red cedar at The Hirsel.

40 metres in Europe. It has a straight trunk and an 'upright' appearance and this one, although growing close to other trees, is not crowded out, with enough space to show off its shape, including its conically shaped top or crown. The leaves are sharply pointed and scale-like and have a very spicy smell when crushed, so much so that cedar leaf oil is used in the production of perfumes, insecticides and other medicinal products. The tree shown must be over 100 years old but it has a long way to go because it is another long-lived species and survives for up to 1400 years or so in America and Canada.

The Western Red cedar is planted for ornamental purposes and also grown worldwide for its timber. The wood has an attractive colour and was used for the cowl in the Welsh Assembly building. It is resistant to decay as it contains the chemical, *thujaplicin*, a natural fungicide. It is therefore used for roof shingles, fence posts, piling, containers and exterior cladding in North America and elsewhere. Native Americans used the wood for building, carving of boxes, totem poles and tools and making canoes, aided by its light weight and durability; its bark was also used by these people for slow matches, baskets, mats and even women's clothing. The tree at The Hirsel appealed to me as a notable specimen of a large, impressive tree that is not only ornamental and useful but was so important to the aboriginal people of the Pacific coast of North America.

On Sir Alec Douglas-Home's 70th birthday in 1973, the Foreign and Commonwealth Office presented him with 105 trees representing all the countries that he had visited. These were planted around the house, the stables and the walled garden (see www.historic-scotland.gov.uk 'An Inventory of Gardens and Designed Landscapes in Scotland'). The gift included a wide variety

Wild pear tree at The Hirsel.

of species that are interesting or unusual. For example, instead of English oak, there is a Lebanon oak and a Caucasian (or Persian) oak and, instead of a Common beech, there is an Oriental beech. There are

42

hawthorns and a rowan tree, but there is also an American sweetgum (*liquidambar styraciflua*, a deciduous tree from eastern North America), a clammy locust (*Robinia viscosa*, a small, deciduous tree of the pea family from North Carolina), a Père David's maple (*Acer davidii*, a deciduous snakebark maple from China) and a Chinese necklace poplar (*Populus lasiocarpa*, a medium-sized deciduous tree from central China). One group of 'birthday' trees can be seen on the grass area outside the west wall of the walled garden and this includes a dogwood (*Cornus macrophylla*, a small deciduous tree from East Asia), a snow-drop tree (*Halesia carolina*, a small, deciduous tree from south-east America with snowdrop-shaped flowers), a wild pear (*Pyrus amygdaliformis*) and the above mentioned Chinese necklace poplar. There is also a tulip tree (not the one in chapter 5 on 'ancient' trees), a birch and a whitebeam (*Sorbus aria*, a small, deciduous, native tree from the rose family).

The wild pear in the photograph on the preceding page is one of this notable 'grass group' of trees. It is also known as the Almond-leaf pear and is a medium-sized, deciduous tree from southern Europe and Asia Minor which, like the whitebeam, belongs to the rose family. The tree is outside the west wall of the walled garden, towards the southern end (Grid Ref. NT82854045). It only has a girth of 1.61 metres at 1.50 metres above ground level. *Pyrus amygdaliformis* is one of many varieties of pear and is a particularly attractive one that is ideal as an ornamental tree, having attractive foliage and a very good shape. It produces a large number of small, tart, apple-shaped pears and, like all pear trees, has a very dense wood. Because it is such a hard wood, pear is ideal for high definition carving especially as it has attractive colour variations and does not have a pronounced grain.

However, it is usually the Common pear (*Pyrus communis*, the European pear) that is used for this purpose.

By contrast the next tree is a lime stump. Why would anyone want to look at a stump? I have included it partly because of its size

The lime stump near The Hirsel house.

(5.06 metres girth), partly because of its position in sight of The Hirsel house and also because it is still alive and throwing out new shoots—an unintended coppicing of an old tree. It must have been a very large tree easily visible from the windows of the house and could well have been old enough to be at least a 'veteran' tree. Although the stump can be easily viewed (at Grid Ref. NT82924061) from the path traversing the site of the medieval (and earlier) village, it is situated off the path on private land just within the policies of the house. A tree that was several hundred years old is showing a vigorous new growth.

Two deciduous trees whose unusual appearance contrasts so markedly with almost every other tree in the landscape can be seen just over the southern fence boundary of Home Park in Coldstream (see next page) and, although they stand on private land, they are close to the fence and tower above it (Grid Ref. NT84053974). These trees are Lombardy poplars (*Populus nigra* var. *Italica*) and, like all poplars, are members of the willow family. They are quite short-lived, perhaps 60 or 70 years but they grow quickly, reaching 40 to 45 metres although the average is nearer 30 metres in height. Their speed of growth makes them useful for planting as windbreaks and they also have attractions for ornamental planting because they quickly provide such a contrast with other trees in the landscape. Their branches start close to the base of the trunk and grow almost vertically giving a tall, columnar appearance to the tree. In England and on the European continent, Lombardy poplars can be seen planted about two metres apart in avenues of trees forming massive windbreaks; they are more common on the continent and they were only introduced to Britain in the middle of the eighteenth century.

Two Lombardy poplars, on the other side of the Home Park fence; Coldstream Community Centre tower, behind.

There are only two native British poplars, the Black poplar (*Populus nigra*) and the aspen (*Populus tremula*), neither of which resembles the Lombardy poplar. The latter is not always favoured by landscape gardeners, partly because of its alien appearance, but also because its roots are very invasive, even after the tree has been felled. It is also susceptible to disease and insect attack; both these can reduce its lifespan. The main use of the Lombardy poplar is ornamental but the timber is hard but lightweight and can be used for making pallets, cheap plywood and packing cases. It can also be used as an energy crop in biomass or biofuel energy systems, because of its fast growth, high yield and carbon mitigation properties.

There are many 'notable' Common beech trees (*Fagus sylvatica*) to be found in and around Coldstream and I will mention just one of them as an example. Beech is of the same family as oak but split off from it many millions of years ago. It grows throughout Europe and spread to southern England after the last Ice Age and is regarded in England and Wales as a native tree. The tree grows well in Scotland but its manner of arrival has always been problematic and it is regarded by some people as an introduced species. If so, it arrived a long time ago and has become accepted as one of our most common woodland trees and it is also a common sight in hedgerows.

Beech requires a moist but well-drained site but is happy with most types of soil and can therefore flourish on high ground in poorer conditions as well as in richer soils in valleys, growing up to 40 metres in height. It is deep-rooted and regarded as a shade-loving tree. Its shape depends upon its location; in a dense woodland environment it grows up straight towards the top of the canopy with its branches at higher level; if given space, it is bushier and branches out at lower

Beech tree outside the boundary wall of the estate offices, The Hirsel.

levels. The beech is happy in both situations, as a solitary tree or when growing in groups, either as pure beech woodland or in a mixed broadleaved woodland but it is vulnerable to drought, flooding and hard frosts. Like the oak, beech does not produce flowers or seeds (beech mast) until the tree is at least forty years old and, even then, seeds are not produced every year. In contrast to the brown, craggy, ridged oak bark, beech bark is very attractive, smooth and silver grey. The lime-green of its new leaves in spring and russet in autumn, probably present the most attractive foliage of all the deciduous trees.

Beech trees can live for up to 250 years and, on that basis, the tree shown on the preceding page is certainly a 'notable' tree and perhaps even a 'veteran' tree, as it is a mature specimen of perhaps 180 to 200 years, having a girth of 5.10 metres at 1.50 metres above ground level. It can be seen on the grass area to the north of the boundary wall of the estate offices at The Hirsel (Grid Ref. NT82814057).

Notable or veteran, it is certainly notable on account of its  size and also because of its twisted trunk and its 'buttress' roots. It is difficult to say whether the tree has a single trunk (known as a 'maiden' stem) that has become twisted due to a quirk in growth, or whether it is a multi-stemmed tree whose stems have fused together during growth (see above). Multi-stemmed trees can occur as a result of early coppicing or from the growth of several seeds that have been deposited close together. Whichever it is, the tree has an attractive trunk and dense foliage. The trunk divides higher up and the shape of the spreading branches towards the top (see preceding page) suggests that the tree has lost an earlier crown at some time in the past, probably due to storm damage. As a harvested tree, beech has many uses because it is such a light-coloured, clean and odourless wood. As such, it is hygienic and ideal for kitchen worktops and children's toys and, in the past, it was used for making tool handles, buckets, bowls and plates. It is also hardwearing which makes it useful for flooring and for constructing workbenches. Beech can be easily worked, even across the grain and is a good timber for making furniture.

Although, as mentioned, beech is vulnerable to unfavourable weather conditions, it is a hardy tree and a survivor. Even if uprooted by storms, it can often throw out new shoots from the old roots. If 'wounded' it has the ability to isolate the wound from the rest of the tree by forming a cankerous growth. Because beech is shade-loving and creates shade because of its tall stands, saplings of other species find it difficult to grow in its shadow.

All trees have their own character but probably oak and beech are the two species that are most familiar to people. Oak is seen as large, craggy, long-living, symbolic of national spirit and the Royal Navy, 'hearts of oak', acorns, oak apples and historic events. It is also an attractive tree and is a familiar inhabitant of our woods. Beech is just as familiar and all-beech woods have a special appeal with their attractive, smooth, silver-grey, columnar trunks, the spring and autumn colours of their foliage, the rustling carpet of dry fallen leaves and the evocative smell of that carpet when wet.

4

Some Veteran Trees

The beech tree in the previous chapter had a twisted *trunk*. The *bark* of the Sweet chestnut tree (*Castanea sativa*), not the trunk, also twists; as the tree ages, it splits into wide, spiralling grooves. The tree grows to a height of about 30 metres and can live for 1000 years or more in its native habitat of southern Europe, Turkey and North Africa, but also to a great age in Britain. It is also known as the Spanish chestnut and belongs to the same family as beech and oak. The Sweet chestnut may have been introduced to England by the Romans, who used the chestnut fruits to make polenta. Polenta is a type of porridge that is still made today under different names in Europe, Africa and North and South America, usually from maize. The fruit of the Sweet chestnut has many culinary uses and should not be confused with the non-edible Horse chestnut. It grows throughout the British Isles, most commonly in the south-east of England.

The Sweet chestnut tree shown on the next page is on the right hand side of the drive to The Lees (Grid Ref. NT83972925). However, it is within the private policies of the house just beyond the PRIVATE sign. It is a large and very old tree with a girth of 6.28 metres at 1.50 metres above ground and, although it has suffered much damage over the years, it is holding its own. The tree's maturity is also evident in the holes, hollow branches and the amount of dead wood in its branches and in the crown. One of the dead branches is even touching the ground as though it might have tried to take root at some time. If it had, it would have shown how some trees can 'walk'

Sweet chestnut at The Lees.

by producing new shoots from the rooted branch and allowing the old tree roots to die over a period of years. This could happen several times over, allowing the original tree to move sideways in several steps. Alternatively, the original tree might continue and just add new trees to sprout around it, forming a small single-species copse.

Sweet chestnuts were planted for landscaping in the 18th & 19th centuries in parkland and on large estates, where some specimens have been recorded with girths of 10 or 11 metres or more, such as the 'Tortworth chestnut' in Gloucestershire which is probably the oldest specimen in Britain, 1200 years old and with a girth of 11 metres. It too has produced new trees around it. The oldest known Sweet chestnut tree in the world is the 'Tree of a Hundred Horses' near Mt. Etna in Sicily which has a girth of over 60 metres and is reputed to be between 2,000 and 4,000 years old. The very large girth is due to the fact that the single root system has developed into many multi-stems that have formed a circle around a space in the middle. The name comes from a legend that Giovanna of Aragon (who became Queen of Naples), together with 100 knights, is said to have sheltered under the tree during a ferocious thunderstorm in the 15th century. The tree at The Lees, although a mere stripling compared with the Sicilian tree, is large, old and mature enough (compared with the Tortworth chestnut) to be on the verge of being considered an 'ancient' tree, but perhaps not quite.

Sweet chestnut responds well to coppicing and, in the past, was coppiced in large quantities in south-east England for charcoal manufacture, which was used in metal working. Kent and Sussex are the major areas for chestnut coppice today and thousands of acres are managed commercially. Young chestnut wood is hard, strong and rich in tannins that make it very durable and suitable for outdoor use. It is still used for stakes, gateposts, and fencing and for cladding on buildings. The wood is light in colour and used for furniture making and, in Italy, is used to make barrels for ageing balsamic vinegar. Tannin from the bark is used for tanning leather in Europe and the UK imports supplies from Italy. The wood provides a 'spitting' fuel for open fires.

In winter, street vendors in Britain used to sell roasted Sweet chestnut fruit (nuts) from wheeled carts, an uncommon sight today. When I was a boy, they were roasted on the fender of the open fire. It is a pity that, although they are eaten in the form of poultry stuffing and other constituents of traditional Christmas fare, their culinary use has declined in recent decades. Unlike most commercial nuts that contain relatively large amounts of protein, Sweet chestnuts contain mostly starch with small amounts of fat and protein. Because of the high starch content, ground Sweet chestnuts were once used to make starch to whiten linen. Also in the past, infusions of the long, toothed leaves and the bark were used for various medicinal purposes.

The next 'veteran' tree that I have selected, is on a fence line between two fields. Before the eighteenth century there were few *dykes* (walls), fences or hedges around fields in Scotland. There were some walls, but these enclosed the immediate policies of country houses to prevent livestock from straying in to them and to protect the privacy of the landowners. At The Hirsel and The Lees and at many country houses, this was in the form of a *ha-ha*, which is a trench or a vertical drop, the nearside of which is a wall, concealed when viewed from the house and designed to give an uninterrupted vista from the garden to the fields or parkland. The garden appears to merge

54

seamlessly into the countryside.

By the middle of the nineteenth century, field boundaries were in place, brought about by *enclosure*, a process that was part of a change in agricultural techniques known as *Improvement*. Before Improvement, the Scottish farm (which was often on a short-term lease from the landowner—many of whom were large estate owners) would have been divided into *infield* (or *in-bye*) and *outfield* (or *out-bye*). The infield was the land immediately around a small group of farm settlements comprising a *fermtoun* and families would farm the infield strips in common on a rotation, with one part fallow every third year. The infield received most of the manure from the animals and was used for growing say, barley and oats. The outfield lay beyond the settlement and was poorer in fertility and drainage; it was also farmed in common. Cattle might be *folded* (grazed) on the outfield before oats were grown for two years, followed by another two years when the crop was left as straw. No further manure was applied and when yields declined, the outfield was left fallow for five years or even more. Beyond the outfield lay moorland or wasteland again used for communal grazing.

The strip system of ridges and furrows, known as *run-rig*, was seen to be inefficient because strips were managed by different individuals and the unproductive furrows between the ridges were wasteful, with inexact boundaries leading to disputes. Also, it was not possible to use new farming methods and machinery. By the authority of many individual Inclosure (sic) Acts during the eighteenth century, landowners were empowered to enclose common lands and open fields with hedges, fences or walls. A Consolidation Act was passed in 1801 and a General Inclosure Act in 1845 that meant that

Commissioners could agree to enclosure without individual parliamentary approval. The commoners lost their rights but enclosure meant that fences prevented livestock from trampling on arable fields and eating the crops. Separate fields could be farmed to take advantage of different soil conditions and hedges or walls could provide shelter. Enclosure, together with the introduction of improved farming methods, improved agricultural production and financial reward. In 1835, *The New Statistical Account of Scotland VI*, pp. 209/210 states that, in Coldstream:

> the farmers are almost all men of capital and enterprise, and possess great skill in their calling: The land is therefore in a high state of cultivation, and great improvements have been introduced of late years, particularly in draining and enriching the soil...The general duration of leases is nineteen or twenty-one years, and the rents are, I believe, in almost every instance paid in money. The lands are well-enclosed...

On page 214, the *Account* states that, 'the system of agriculture pursued in this parish is not inferior to that of the best districts of Scotland...' This was a good endorsement of agricultural practice in Coldstream parish. The present day size of the oak tree on the field boundary (see next page) shows that the farms on The Hirsel estate must have been enclosed quite early on, because this is a large tree and one of an evenly spaced row of large oak trees that must be well over 200 years old.

Field boundaries in the Borders were stone dykes, hedges and, probably later on, fences. Stone dykes are not endemic to Coldstream and the surrounding area, except for house and garden boundary walls because the rich soil, farmed over the centuries, did not produce surface gatherings in sufficient quantity. This contrasts with higher farms north of the Tweed valley, where stone walls were often built from choice and still survive, for example, on either side of the A697 north-westwards from Greenlaw to Carfraemill.

Oak tree at the end of a line of field boundary oaks, The Hirsel.

The boundary oak (see also view from the other side, below) is sited along a present day fence, which may have superseded a hedge. It is on land that is not open for public access but can be seen from the bridleway and path along the south side of The Hirsel lake (Grid Ref. NT82364001). It is a substantial tree, having a girth of 5.68 metres at 1.00 metre above ground level (below the normal measuring

height because of the large branch, see left) and stands at the end of a line of evenly spaced mature oak trees that were probably planted during the eighteenth century. Is this one really a 'veteran' tree—or just a good, substantial, mature and 'notable'

specimen? I am not sure, but I am not too worried about any strict categorization if the tree stands out for me. Oaks planted in hedgerows provided shade but they often had a special use by being *in situ* for providing the best material for making nearby farm gates.

The oak has a fascination all of its own. Apart from its usefulness as a traditional building material and the historical associations mentioned earlier, acorns and oak apples have always been attractive to children and are useful in their own right. Acorns are a valuable part of the diet of pigs (and wild boar in medieval times), deer and squirrels as they are a useful source of protein, carbohydrate and fat. In the past, some cultures have used acorns for roasting and for making acorn flour and in World War II, they were

ground and roasted for use as an ersatz coffee. As a small boy, I used to collect green acorns and peel off the skin to reveal the two halves of the nut. The separated halves were hollowed out into miniature rowing boats for floating in the bath or in the wash hand basin. Oak apples also had childhood appeal but they were really too light to be used as marbles. Their name derives from their apple-like shape, although their diameter is

only about 2 cm. and they are brown. They are not fruits, but galls that are formed when a female oak apple gall wasp injects an egg into the mid rib of a newly grown leaf. The larva hatches inside the leaf and a chemical reaction causes the leaf

to form into a woody gall around the larva which pupates and drills its way out of the gall as an adult wasp, leaving an exit hole (see above).

In medieval times, oak apples were used for making ink, as they contain tannins. The best ink was made by crushing the galls into powder, boiling in water and allowing the liquid to ferment until a mould formed which converted the *gallotannic acid* into *gallic acid*. The mould was skimmed off and *ferrous sulphate* added to the fluid, together with some *gum arabic* (which is the sap from acacia trees in North Africa) for thickening purposes so that the ink could be used with a quill pen. The ink was not very dark when first applied to paper but would turn black eventually from exposure to air. To get round this problem, natural plant colourants were often used to inject immediate colour; *indigo* was also included as it acted as a preservative.

There are three Common lime trees along a path which forms part of the 'Riverside Walk' (colour coded 'yellow') at The Hirsel, on the east side of a field next to the River Leet. They have survived previous forestry thinnings in this woodland area and remain as mature and very impressive trees framing the outer edge of the path.

Mature lime tree, No. 1, Riverside Walk ('yellow walk'), The Hirsel.

The tree on the previous page (Grid Ref. NT83134061) has a girth of 5.30 metres at 1.50 metres above ground level. The tree below (Grid Ref. NT83184057) has a girth of 7.50 metres at a height of 1.00 metre above ground level. This tree (No. 2) is multi-stemmed and has a great deal of epicormic growth making it necessary to measure the girth at a lower level, threading the tape inside the tangled growth.

Mature lime tree No. 2, Riverside Walk ('yellow walk'), The Hirsel.

The third lime tree (Grid Ref. NT83254051) has a girth of 4.85 metres at a height of 1.50 metres above ground level. This tree and perhaps trees 1 and 2 as well, could be multi-stemmed trees where the trunks have fused together in the past, whilst remaining separate at higher levels.

Mature lime tree No. 3, Riverside Walk ('yellow walk'), The Hirsel.

On the bank of the River Tweed on the side of the great loop in the river before it passes Coldstream, is what I think is an old, battered looking Crack willow tree (*Salix fragilis*, see below). It stands in splendid isolation in a damp environment next to the sign for

Crack? willow on the bank of the River Tweed next to the Bags pool.

the Bags pool of The Lees fishing beat (Grid Ref. NT84973924). It may have been pollarded very early in its life but, whatever its early history, a large branch or even one of its trunks, has been sawn off at

some time and there has been damage to many of its branches with a great deal of dead wood lying around, seen left in this view taken in winter. At one time, it must have had a companion a few metres away, now reduced to a stump. Crack willows (if this is what it is) do not tend to live to a great age. Because of their deep root system and fondness for wet places, they are often planted along rivers to bind and protect the banks. They grow quickly because of the damp soil but then become top heavy with the crown often leaning over, causing great cracks to appear in trunk and branches. This, in turn, allows rainwater penetration, the entry of diseases and the eventual destruction of the tree.

The appearance of this particular tree led me to classify it as a 'veteran' tree because even though it cannot match the age of other broadleaved deciduous trees, it looks as though it has reached that stage within its shorter lifecycle. Despite the damage to its trunk and branches, it has a vigorous growth and looks set to carry on for many more years. It is a substantial tree with a girth of 4.92 metres at 1.50 metres above ground level.

The Crack willow is difficult to distinguish from the White willow (*Salix alba*); both have narrow pointed leaves but the Crack willow leaf is hairless whilst the White willow leaf has short downy hair. The White willow can be a taller tree but it is quite common for hybrids to form from cross-fertilization, which confuses identification. Willows fall into three categories, the primary group being the Crack willow, the White willow and the Weeping willow (*Salix babylonica*). Then there are the hedgerow willows known as *sallows*, which are smaller trees with broader leaves and finally, the small, shrubby *osiers* that grow in very wet conditions. The featured

Crack? willow stands on the river bank close to the long earth and grass *bund* built by Napoleonic prisoners of war around the fields within the loop of the river (see left) opposite the Baa Green. There is also a bund that crosses the fields. The signboard at the start of the riverside footpath states that the construction of the bunds was commissioned by Sir James Marjoribanks in 1820, although it was Sir John Marjoribanks, 1st baronet of Lees who held the estate at that time and until 1833 when he was succeeded by his son, William, who died in 1834. After Napoleon's defeat at Waterloo in 1815, French prisoners returned home, but whatever the date and whoever commissioned the work, it was a colossal task. Prisoners had originally been kept in hulks on the River Thames but, as the war progressed and as Wellington achieved success in the Iberian peninsula, more and more prisoners had to be accommodated in prisons such as Dartmoor which was built for the purpose. Prisoners were sent to Scotland, initially in central Scotland, including Edinburgh Castle but eventually many were dispersed to Border towns. Officers were held separately and were able to live a fairly free existence in many small towns subject to parole agreements. Audrey Mitchell in *Historic Kelso*, p. 57, mentions that 16 Horsemarket was originally Kelso's theatre, built by French prisoners (I think officers) during the Napoleonic wars.

The fields at this point on the River Tweed are subject to

regular flooding and the rich silt deposited on the land was good for improving the fertility and soil composition in arable fields. The bunds or dykes were built as a very practical way of controlling the retention or release of floodwater using a number of sluices, such as this one

(see above) in the side of the bund, below the willow tree.

66

Willows have the ability to reproduce themselves not only from seed but also from stem rooting. With the Crack willow, this can happen when branches split and fall to the ground or are swept down river; there, they take root, grow quickly and form new trees. New growth also sprouts from stumps or even from felled timber. Willow must be the easiest tree to propagate by pushing even a small stem into the ground where it will quickly form roots.

The wood from all the species of willow has had many uses down the centuries because of its inherent properties but also because it can be easily harvested from fast growing coppiced or pollarded trees. The timber is tough but lightweight and has been used for building and sometimes boat building (particularly coracles), for planking, cleft post and rail fencing and for making charcoal. Thinner poles were made into baskets, trugs, eel traps, and hurdles; cleft strips could even be twisted into rope. Cricket bats are made from a White willow hybrid (*Salix alba* var. *caerulea*) which has qualities of lightness, durability and light colour that are ideal for the purpose. The bark of the White willow contains *salicin* which was used in the original manufacture of *aspirin*.

Back on The Hirsel estate, a large old beech caught my eye. It does not stand alone but in a belt of mixed woodland to the east of the walled garden (Grid Ref. NT82994045). However, it stands out for me even though it is amongst other trees and close to an 'ancient' tree which I will describe in the next chapter. The beech must be about 200 years old and could be described as multi-stemmed, rather than having a maiden trunk. The tree was perhaps pollarded in its early years but, whatever its history, it gives an impression of gnarled old age. One of the branches has crossed and fused with one of the main

Old Common beech tree in winter, with The Hirsel house behind.

stems and there are also a number of *burrs*; these are woody growths on the trunks, formed by the tree to cover wounds.

      The name for beech is Germanic in origin, probably from the

Old English, *boc* meaning 'book' because beech bark was a common writing material before the development of paper. The botanical name, *fagus*, is from the Greek 'to eat', which refers to the beech nuts or 'beech mast'. The nuts are very small and difficult to shell from their angular shells. As a child, I never seemed to be able to extract a decent sized nut and they do seem to vary in quality. They also have to be found before the squirrels get to them but, if successfully harvested, they have a high fat content and contain an oil which, in some parts of Europe, has provided a good quality alternative to olive oil. They are useful food for deer and served as foraging for pigs or wild boar in medieval times.

One of my favourite veteran trees is one that has even more burrs on its trunk than the old beech tree. It is an old pollarded sycamore (see next page) which stands in the middle of a pasture field at The Hirsel; because of this, it cannot be approached because it is not on a public right of way, although it can be seen at a distance from the Riverside Walk (the 'yellow walk') to the east of the River Leet (Grid Ref. NT83154051). The advantage of having a tree in the middle of a pasture is that it provides shade for cattle in hot weather— the additional advantage of a sycamore is that it forms a natural dome and has big leaves that increase the amount of shade.

In this field, it is particularly useful because this is one of the frequent summer pastures for the Douglas and Angus Estates' fold of Highland cattle. The many burrs may have been partly caused over many years by cattle brushing against the tree or grazing on side shoots and causing damage that has been successfully repaired by the tree in the form of growth over the wounds. They may also have formed when the tree was pollarded during its early life.

The tree has a substantial girth of 5.67 metres but the measurement had to be taken at a height of only 1.10 metres above ground level because of the distorting effect that the burrs would have had on the measurement at the normal height.

Old pollarded sycamore in the pasture field at The Hirsel.

A closer view of the burrs on the sycamore trunk and branches.

The sycamore is suited to an unsheltered position in a field because it can withstand the cold and it will maintain its symmetrical appearance in windy positions. It can also grow in all kinds of soil.

71

There are four yew trees immediately north of the obelisk on The Hirsel estate. They are over 200 years old but they are not ancient as yew trees go, if they are compared with the oldest known specimens in Britain, the Llangernyw yew near Conwy, North Wales and the Fortingall yew in Perthshire. The former may be 4,000 to 5,000 years old and the latter, which has a girth of almost 16 metres, anywhere between 2,000 and 5,000 years old, although Archie Miles in *Silva,* Ebury Press, 1999, has mentioned the possibility of 9,000 years for the Fortingall yew. If this figure were true, it would make it the oldest living tree in Europe, beaten only by a living root system (but not a standing tree) of a Norway spruce (9,550 years) discovered in Sweden in 2008 (*National Geographic News*, April 14, 2008). Both yew trees are difficult to age because they exist in separate fragments around large empty circles, many parts having been cut away in the past for souvenir pieces or souvenir carved bowls.

There are other yews on The Hirsel estate, some of which are of the same age and some much younger trees. This group of four is not even a good representation of the species because of the extensive pruning that has been carried out in recent years to remove heavily decayed limbs. At 227 years, they are not 'ancient', not really 'veteran' and might not even be considered 'notable' by many, but they are old and veteran in my eyes and I am including them here because they have historical significance. They stand immediately to the north of the obelisk erected by the 9th Earl of Home in 1784 in memory of his elder son, William, Lord Dunglass who, as a Lieutenant in the Coldstream Guards, died of wounds, aged 24, after the battle of Guilford Courthouse in December 1781 in the American War of Independence. The British were victorious but suffered such heavy casualties that it was regarded as a *pyrrhic* victory.

Obelisk erected by Reverend Alexander Home, 9th Earl of Home in memory of his son, Lieutenant William Home, Lord Dunglass, of the Coldstream Guards who died of his wounds in December 1781 after the battle of Guilford Courthouse in the American War of Independence.

The four trees are situated such that they must have been planted at the same time as the construction of the impressive obelisk. Unfortunately, the trees present a sorry appearance because of the drastic pruning and two may be dead. The trees and the obelisk can be

The four yew trees, two living, above and two possibly dead, below.

seen less than half a mile north-west of Dunglass Bridge overlooking the River Leet (Grid Ref. NT82014135). The four trees have girths of between 2.10 metres and 3.75 metres, taken at ground level.

The Common yew (*Taxus baccata*) is an evergreen conifer, native to Britain, with flat needle leaves that are highly toxic, as are the seeds inside the bright, red berries. It is a small to medium sized tree growing to 10 to 20 metres in height. Yew trees are often associated with churchyards and it may be that they were planted over the graves of plague victims to protect and purify the dead or perhaps to sanctify any Christian burial. A great deal of mythology has grown up over the centuries and, although yew trees are symbols of immortality in many traditions, they are also seen as omens of doom. The yew grows well in many conditions, either singly or in groves. It tolerates shady conditions and it may be that its dark colour and sometimes gloomy growing conditions, gave rise to mystical thoughts and beliefs. Many of these beliefs were positive, however, which is the reason why it was so often planted in churchyards, gardens and elsewhere as a force for protection and good. This was undoubtedly the reason why the trees were planted next to the obelisk.

Yew wood is very hard and heavy but elastic, making it suitable in the past for making longbows and spears. Because it is so hard, it was also useful for making shuttles, cogs, axle-trees and pulleys. The heartwood is red and the sapwood is white, the attractive combination making it ideal for laminating furniture and for making musical instruments, bowls, ladles or other objects that benefit from an artistic finish. The drug, *docetaxel*, is extracted from the needle clippings of cultivated species of the Common yew and is used as part of the process of making anti-cancer drugs.

5

Some Ancient Trees

'Fat ones, thin ones, gnarly and hollow ones'

So goes one of the descriptions used by the Woodland Trust as a catchphrase for their nationwide 'Ancient Tree Hunt' which I mentioned in the Introduction. It gives a good mental picture of the wide range of trees to be recorded for the hunt or even for inclusion in this book. The first, third and fourth adjectives are the ones that fit most comfortably with the description of ancient trees, or even some veteran trees (see the distinctions mentioned at the beginning of Chapter 3). But, the second one, 'thin', can also be applied because an ancient oak growing out of a soil-less rock face and contending with harsh climatic conditions will often have a thin and stunted look.

I have indicated in previous chapters that I have often found it difficult to place many trees into one of the three categories but I have also suggested that it doesn't bother me too much. A tree is worth noting for itself—and that has been the basis of my approach even if it does not tick all the boxes. In this chapter I have described four trees that, to me, are 'ancient' even if, for example, I cannot say that they are in the third and final stage of their lifespan. They are all on The Hirsel estate but there will be other trees elsewhere in the Coldstream area that I have yet to discover and are well known to other people.

A hollow-trunked, ancient ash tree at The Hirsel.

The first of these is an ancient Common ash (*Fraxinus excelsior*) next to the golf course on the path east of The Homestead Visitor Centre (Grid Ref. NT82894020). It has a girth of 4.75 metres at 1.50 metres above ground level. It has a hollow trunk visible only from one side (see below) with holes in the branches, dead wood in the crown and dead wood lying on the ground. There has been much damage but the tree is still thriving and there is a good canopy.

It is an interesting fact that when a tree trunk becomes hollow, it does not necessarily become weakened but can become stronger or more stable. As a large tree matures, sap is no longer taken upwards by the heartwood; this is done by new growth formed every year on the outside of the trunk (in a cross section of a tree, the

outermost annual rings). The heartwood becomes more or less dead tissue and, over many years, fungi rots it away leaving a large, old, hollow tree trunk. The outside layers or rings of sapwood continue to supply all the nutrition needed by the tree. Although the tree has lost support from the heartwood, this is more than off-set by the fact that the tree is now much lighter and has less weight to hold up, whilst the strength increases in the growing sapwood. At the same time, a wide trunk is harder for the wind to bend than a young, slim one and the hollow trunk provides less resistance as the wind may whistle through it. Whereas in the past it was thought that a hollow tree was unstable and had to be cut down, it is now realized that young, tall and slender trees are much more likely to be blown down by high winds than old, thick, hollow trees (although I have to admit that the appearance of the ash trunk, above, might suggest otherwise).

78

Sometimes, landowners have not appreciated this and have poured concrete into a hollow trunk and/or filled it with bricks as has happened in the case of an old, hollow beech (see right). Unfortunately, the infill can shorten the life of the tree rather than extend it. The fill material will expand and contract at a different rate from the living tissue and can create problems of its own with enclosed damp penetration leading to disease, or causing damage to the outer growth layers  from the way in which it is inserted into the hollow. However well it is done, it is unlikely to add any strength to the trunk.

Ash is perhaps the third most well known tree in British woodlands after oak and beech. But this is a subjective view and other people may have other ideas, perhaps putting forward the sycamore or the Silver birch or the Horse chestnut as trees that are just as well known to everyone. However, I think that I can say that deciduous, broadleaved trees are more familiar than conifers because, perhaps with the exception of the deciduous larch, other conifers can be quite difficult to identify.

Ash and oak both have roughly textured bark although ash is grey in colour, whereas oak has a grey-green-brown bark. Beech has a smooth, grey trunk that is quite different from oak and ash; all three trees have easily distinguishable leaf shapes as seen in the illustrations on the next page and ash has distinctive black buds compared with the long, thin, brown, beech buds and the stubby, brown, oak buds.

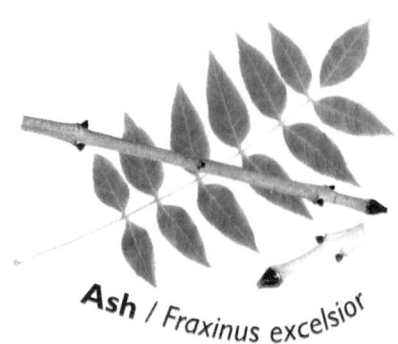

Ash, beech and oak leaves and twigs; three quite different leaf structures.

(Reproduced by kind permission of The Woodland Trust).

The ash leaf is a *compound* leaf (with several leaflets growing from the same stem) and *pinnate* (where leaflets are paired). The beech and the oak are both undivided, *simple* leaves but, whereas the oak has a definite *sinuate* or undulating margin, the beech only has a slightly wavy margin.

Ash is a deciduous, broadleaved, hardwood and is native to Britain. The name 'ash' may come from the Anglo-Saxon, *aesc*, meaning 'spear' because the hard, dense and flexible wood was so useful for making such weapons. *Fraxinus* is the Latin for 'ash' and *excelsior* comes from the Latin for 'high' or 'lofty'. William Gilpin in his *Remarks on Forest Scenery and Other Woodland Views*, ed. Sir Thomas Dick Lauder, 1834, Fraser & Co., Edinburgh, Vol. 1, p. 81, says that 'I have sometimes heard the oak called the Hercules of the forest; and the ash, the Venus'. He compares the oak which 'joins the idea of strength to beauty' with the ash which 'joins the ideas of beauty and elegance'.

Comparison can be made between oak and ash in other ways. The ash usually comes into leaf much later than other native British trees and usually drops its leaves earlier and faster, without producing any autumn colouring. The timing of its spring growth compared with the oak has given rise to the old weather forecasting rhyme 'Oak before the ash and there'll be a splash; ash before the oak and there'll be a soak'.

Ash is not a very long-lived tree, with few living more than 200 years and most trees for less than that, although Forestry Commission Scotland believes that the Gordon Castle ash at Fochabers, Morayshire, which has a girth of 7.79 metres, dates from about 1769. The tree at The Hirsel, although perhaps younger, shows by its size and appearance that it is an 'ancient' tree even though it is the youngest of the trees mentioned in this chapter. Ash has many uses, not least its excellent capacity for burning even when the wood is 'green'. This may be reflected in the Latin name, *Fraxinus* which is associated with firelight. The tough and elastic properties of ash mean that it can take heavy weights and it was used for coach axles (and sometimes the spokes and felloes [wooden rims]), oars, tool handles and weapons such as spears or halberds. The wood can be used for making furniture and coppiced poles are ideal for use as hedging stakes, bean poles and tree stakes; the wood is often used for sporting equipment such as cricket bats, hockey sticks and snooker cues. The elasticity of the wood made it ideal for making longbows and, more recently, the flexible chassis frames of Morgan sports cars.

The oldest oak tree that I have found in Coldstream, so far, is on the east side of The Hirsel on the left hand side of the path before reaching point 4 on Route 4 as described in *Paths around Coldstream*, Scottish Borders Council, 2007 (Grid Ref. NT83164088). The tree is tucked into the woodland and is quite difficult to see at first although it towers above the woodland canopy. It has a girth of 5.70 metres at 1.50 metres above ground level and there is dead wood in the crown and on the ground; it has also lost limbs from storm damage and from pruning. However, the trunk is sound with no hollows and no sign of splitting into separate growing sections like other old trees. This Pedunculate oak (*Quercus robur*) must be over 400 years old, but does that make it an 'ancient' tree? Oaks are noted for their longevity and this one does not compare, say, with the 'Capon Tree' in Jed Forest near Jedburgh which has a girth of 10 metres and is said to be over 1000 years old. There are oaks in Britain that are even older such as the 'Major Oak' in Sherwood Forest (10 metres in 2002), the 'Big Belly Oak' in Savernake Forest (11.20 metres in 1999), the 'Darley Oak' in Cornwall (11.60 metres in 2008) and many others. So, if an oak is said to spend 300 years growing, 300 years resting and 300 years declining, this one is not really 'ancient'. However, it is old and impressive enough for me, so I am including it here, anyway.

Very old oak tree, on The Hirsel estate. Because it is surrounded by other trees, it is not possible to take an 'all in one' photograph. The girth of 5.70 metres was measured above the splayed base and below the healed wounds of former branches, at the standard height of 1.50 metres above ground level.

View showing the thick, ridged bark of the old oak. Note the dead limb rising towards the canopy and the healed wounds from past pruning.

In an earlier chapter I mentioned the demand for oak timber for building and the pressing need for timber to build the ships of the

Royal Navy from the 16th century onwards; this would have been until the first British, iron-hulled, armoured warship, HMS *Warrior*, was built in 1861. The demand for oak even persuaded (reputably) Admiral Collingwood, the commander of one of the lines at Trafalgar (the other being Admiral Lord Nelson) and later, Commander of the Mediterranean Fleet, to carry a pocket full of acorns for planting on his travels, particularly around Northumberland. He planted many oaks on his estate at Collingwood House, near Morpeth.

The old oak at The Hirsel may well have been an ancient pollard judging by the squat trunk with thick boughs originating from a point that is now above head height. If so, this could be the reason why the tree has survived for so long and why its present healthy condition augers well for centuries to come. Pollarded oaks are either spared from felling or they live longer anyway. Firstly, pollarding produces short, stumpy growth that is of little interest to foresters seeking long lengths of building timber or timber for ship construction. Secondly, the shorter trunks and boughs with less foliage mean less weight, better resistance to wind damage and less stress on the root system, giving the pollarded tree a greater natural life expectancy. A non-pollarded oak may only last for 200 or 300 years, its life cut short for natural or unnatural reasons.

In the middle of the walled garden at The Hirsel (Grid Ref. NT82924049) stands a remarkable tree that must be one of the oldest specimens of the tulip tree (*Liriodendron tulipifera*) to survive in Britain. It is a deciduous, broadleaved tree (a member of the Magnolia family) that originates in the eastern parts of North America and can grow up to 60 metres, although only about 35 metres in Europe. Kew Gardens state that a specimen of the tulip tree was brought from America by plant collector, John Tradescant the Younger and is recorded to have been growing in Fulham in 1688. During the 18th and 19th centuries, the tulip tree became popular in parks and large

Tulip tree in walled garden, The Hirsel viewed from gates on south side.

gardens; Kew's oldest specimen dates back to the 1770s and was possibly planted when Charles Bridgeman landscaped Queen Caroline's Richmond Estate (now the western side of Kew Gardens).

The tree has cup-shaped white, green and orange blooms that look like tulips. This is the reason for the name but I also think that the

dark green leaves resemble tulips because of the cut-off apex and the lobed sides of the leaf (see above).

The tree at The Hirsel is within the leased, walled garden and cannot be approached without permission although it can be seen from the 'yellow walk' through the wrought iron gates on the south side. It has a large girth of 8.20 metres, measured at 1.50 metres above ground level. The 25 inch 1st edition of the Ordnance Survey map (1862, surveyed 1858, National Library of Scotland) shows the walled garden divided by paths into four quadrants with presumably the tulip tree standing in a circular area in the middle. Today, most of the garden is down to grass with a small area used for vegetables; the ruined 1870 greenhouses and the newer ones are along the north wall.

The eminent gardener of his day, J. C. Louden, in his *Arboretum et Fruticetum Britannicum or The Trees and Shrubs of Britain*, Longman, Orme, Brown, Green and Longmans, 1838, Vol. 1, pp. 284-291, gives a comprehensive description, geography, history, properties and uses of the tulip tree. He also lists the location, age, height, girth and branch diameter of a considerable number of tulip trees throughout Britain, Ireland and abroad. He lists the oldest tree as being in Fulham which, in 1838, he says is 150 years old with a height of 55ft., a trunk diameter of 3ft. measured at 1ft above the ground and with a branch head which he describes as being in a decayed state, at 25ft. This would have been the tree brought in by Tradescant and was only one of many exotic trees cultivated by Henry Compton, Bishop of London, at Fulham Palace. According to Alicia Amherst in her *London Parks and Gardens*, Chap. 12, 1907, the tree died about 10 years before the publication of her book. Louden also lists the largest tulip tree as one at Syon House, near London which, in 1838, was about 70 years old with a height of 76ft., a girth of 2ft. 6in. and a branch head diameter of 46ft.

Of all the tulip trees that Louden lists in 1838, the one at The Hirsel is the second oldest, at 100 years, but he describes it as a 'low tree' at 13ft. He gives the diameter of the trunk as 4ft and the

branches at 33ft. in diameter. A diameter of 4ft. is equivalent to a circumference or girth of 12.56ft, or 3.828 metres. Some 173 years later with the tree now 270 or more years old, the girth is 8.20 metres and the height is much more than 13ft. It seems that the upper growth was pollarded in the 1960s to reduce the weight and strain borne by the old trunk. There has been vigorous new upper growth and the tree appears to be healthy although showing its age, with dead wood in the crown, holes in the branches and a hollow trunk (see above). The tree may be home to bats and certainly accommodates invertebrates including a wasp *byke* (on the inspection day) up inside the hollow trunk and not visible in the photograph.

Lady Caroline Douglas-Home has seen evidence that the tree was planted in 1742; this would have been during the time of Lt.-Gen. William Home, 8th Earl of Home, who succeeded to the title in 1720 and died in 1761. He gained the rank of Cornet in 1735 in the service of the 2nd Dragoon Guards and held the office of Representative Peer (Scotland) between 1741 and 1761. He attained the rank of Captain-Lieutenant in 1743 in the 3rd Foot Guards and fought in the Battle of Prestonpans in 1745 on the government side. He was Colonel of the 48th Foot in 1750 and of the 29th Foot in 1752 becoming Major-General in 1755 and Lieutenant-General in 1759. He held the office of Governor of Gibraltar between 1757 and 1761.

The charity, Parks and Gardens Data Services, a partnership between the Association of Gardens Trusts and the University of York (see www.parksandgardens.ac.uk), credits the 8th Earl of Home with laying out a formal landscape for The Hirsel comprising two large enclosed parks separated by a double avenue and construction of the walled garden (their Record Id: 1753). The brickwork used confirms this period of construction for me and the probability is that the 8th Earl chose to plant the tulip tree as a feature tree to eventually enhance his walled garden. Historic Scotland, as well as giving a comprehensive history of The Hirsel landscape, also lists the planting date as 1742 (see www.historic-scotland.gov.uk 'An Inventory of Gardens and Designed Landscapes in Scotland'). So, Louden was not far out with his estimate.

The timber of tulip trees is light, soft and easily worked which made it useful for building canoes in Eastern America and for general building purposes, furniture making and as a source of good quality charcoal. Today, it is still used in America but is very expensive. In Britain, it was only planted for ornamental purposes.

The tulip tree at The Hirsel on a training day organized by the Borders Forest Trust for volunteers taking part in the Woodland Trust's 'Ancient Tree Hunt'. This view shows the thick trunk at lower level; above this, the pollarded limbs have replaced by vigorous newer growth.

Just a short distance from the tulip tree, outside the east wall of the walled garden stands an ancient sycamore known locally as 'The Flodden tree' (Grid Ref. NT82944047). It is next to the footpath

The ancient sycamore, the 'Flodden tree', at The Hirsel.

which is on the 'yellow walk' at The Hirsel estate and can be easily seen as the path turns at the south-east corner of the garden. The tree has a short, squat and stout girth with many substantial subsidiary trunks and branches that form a huge dome of foliage. It is difficult to

A closer view in winter showing the massive 'skeleton' of the sycamore.

say whether the main trunk is a single-stemmed, maiden trunk or whether it is a multi-stemmed trunk. The low 'springing' point of the lower branches also suggests that the tree may have been pollarded hundreds of years ago. The girth is 6.80 metres, measured at 1.50 metres above ground level and the height of the tree must be about 30 metres, or 100 ft. The tree is healthy, but its extreme age is evident in the hollow branches, holes and dead wood in the crown and on the ground. Over the centuries, it has attracted insect life and lichens and has grown burrs as scar tissue to cover damage to its limbs. There has been human intervention by pruning and by the installation, a long time ago, of supporting chains and, more recently, some rods to support the limbs. The 'dog collars' holding the chains were cut through in the 1960s but as they had ring-barked the limbs, the advice was to leave them and let them be grown over—as with the nuts on the ends of the rods. Chains were left hanging as part of the history.

The name, 'Flodden tree', arises from the fact that the tree was reputed to have been planted to commemorate the Scottish dead who fell at the battle of Flodden in 1513. It was Alexander, 3rd Lord Home who, with the Earl of Huntly, was in the vanguard of the Scottish army and dispersed that part of the English army that was facing him. He was criticized for then leaving the field without giving assistance to the rest of the Scottish army—this was defeated with great loss of life including James IV and the majority of the Scottish nobility. The Scottish dead may have been as many as eight thousand. The 3rd Earl survived but Sir David Home of Wedderburn and his son, George, one of 'the seven spears of Wedderburn' were both killed. Having turned against John Stewart, Duke of Albany and Regent of Scotland, the 3rd Earl of Home was eventually beheaded in 1516.

In considering whether the sycamore is likely to have been planted at the time, I have mentioned previously (see p. 10) that there is uncertainty as to when the sycamore as a species was introduced to

The 'Flodden tree' in winter.

Britain. The charity, the Royal Forestry Society currently states:

> When and who first introduced sycamore to Britain is
> uncertain. It may have been the Romans but it was still scarce

here in the 16th century and has only really become established over the last 200 years.

(www.rfs.org.uk/learning/sycamore)

Forestry Commission Scotland in *Heritage Trees of Scotland*, pub. Forestry Commission Scotland in association with the Tree Council, 2006, p. 86, states that, until recently, the oldest known sycamore in Scotland and perhaps in the United Kingdom was the Newbattle Abbey sycamore near Dalkeith, thought to have been planted about 1550. Before it was blown down by a freak gust in 2006, it had a girth of 5.486 metres, less than the 6.80 metres of the 'Flodden tree'. However, it had a single-stemmed, 'maiden' trunk and may not have been pollarded during its early years, giving a taller, thinner trunk.

It is possible that The Hirsel sycamore was planted shortly after the battle of Flodden but, if so, it could not have been by the Home family, who did not own The Hirsel at that time. I have heard it said that the tree might have been planted by a descendant, perhaps by the grandson of the 3rd Lord Home. The 3rd Lord Home had no legitimate children and was therefore succeeded by his brother, George Home, 4th Lord Home, who had his titles restored in 1522. He died in 1549 and was succeeded by his son, Alexander Home, the 5th Lord Home who was accused of treason in 1573 and forfeited his titles. His son, Alexander Home, the grandson of the 3rd Lord Home, had his titles restored in 1578 and became the 6th Lord Home and then 1st Earl of Home. He was a loyal servant of James VI. He purchased The Hirsel from Sir John Kerr in 1611 and died in 1619.

If the 'Flodden tree' was planted by the 1st Earl of Home, it would have been after 1611; if the tree was planted before then, it would have been by the Kerr family. One of James IV's grievances before Flodden was that the Scottish Warden of the Middle March, Sir Robert Kerr, had been murdered by John Heron, the bastard, whose half-brother, William Heron held Ford Castle. William was held hostage in Scotland in his place, while John remained free to fight on the English side at Flodden. The Kerrs fought at Flodden under Sir Andrew Kerr of Ferniehurst, who survived the battle. Whoever planted the tree, the sycamore is, hopefully, an enduring reminder of the events of that dreadful battle and a living memorial to the Scots who died on 9th September 1513.

6

Further Thoughts

In this book I have been writing about trees in and around Coldstream that I have selected as being notable for their appearance, their age, or their historical connections. I have mentioned before that it is a subjective and representative selection; other people will have other ideas. In my selection, I have looked at *individual*, notable (for me) trees and readers might well question the exclusion of examples of well-loved 'favourites' such as the Silver birch (*Betula pendula*) with its beautiful bark and shimmering foliage, or the deciduous conifer, the European larch (*Larix decidua*) with its soft, green, tufted needles.

If I had considered say, favourite *species*, I might well have included these and others as well, such as the Horse chestnut (*Aesculus hippocastanum*) with its sticky buds, its large, palmate leaves, its early-flowering 'candle' flowers and its shiny, dark-brown 'conkers', so desired by me as a boy and by many millions of other children. If I had been writing a comprehensive guidebook to all the trees in the Coldstream area, I would have needed to include alder, elder, hazel, holly, hawthorn, Wild cherry (gean), hornbeam, rowan, maple and other native and non-native species, whether notable or otherwise. However, although I did not wish to write a guidebook, I have included a list of trees that are native to Britain and Scotland as an Appendix.

In Chapter 1, I mentioned the growing public and institutional awareness of the benefits of tree cover and the acknowledgement of this in government and local authority planning policies and guidelines. The protection of trees and woodlands and encouragement to supplement tree cover forms part of Scottish Borders Council's adopted Local Plan and the Council's Supplementary Planning Guidance, *Trees and Development*, March 2008, sets out specific requirements for developers in respect of land and tree surveys, development constraints and tree preservation orders. Another Supplementary Guidance, *Landscape and Development*, March 2008, puts landscape issues, including the location of existing trees and proposals for new planting, at the heart of development proposals.

In recent years, government funding has been available for the creation of native woodland as part of the Scottish Rural Development Programme. This has provided grants to farmers and landowners as initial payments for planting on agricultural or abandoned land, including deer or stock fencing and for annual maintenance up to five years. Other grants have been available that tend to come and go, for planting, coppicing or laying of hedges on field boundaries, under the government's Rural Stewardship Scheme. Free hedge plants and financial help are sometimes available from other community orientated charitable organisations.

In and around Coldstream, it is encouraging to see areas of new or recent tree planting with protective tubes to provide shelter belts or replacement of felled areas. Some planting has improved the amenity of an area or has been carried out to improve the stability and

98

Recent planting of trees on former arable land at The Hirsel.

water take-up in areas subject to flooding. Also in the Coldstream area, there are many prominent examples of gaps in hedges having been infilled with double rows of new plants protected by their rabbit-proof tubes. Hedges are really only trees that are planted close together in rows and pruned to height and width requirements. Ancient hedges are not common in Scotland as most hedges were planted after the Inclosure Acts but, from the 18th century onwards, hedges contained different species grown for their timber or for their fruit and nuts as well as those grown just to act as stock or wind barriers. The most common tree used for hedging in recent years has been the Common hawthorn but all hedgerow species provide a valuable habitat for wildlife as well as acting as 'corridors' allowing safe passage through farmland for vulnerable species.

A winter view of a six years old mixed woodland plantation on both sides of the footpath on the floodplain of the River Leet (The Hirsel Golf Clubhouse up on the right).

My further thoughts end on an optimistic note. Figures (see Chapter 1) show that tree cover in Scotland has increased since World War II although there is still a long way to go before reaching current levels in the European Union. There have been worrying trends across the United Kingdom as a whole with figures showing that the rate of new planting halved between 2004 and 2009 despite major woodland initiatives such as the 200 square mile National Forest in the Midlands of England and other schemes. There have also been reductions in woodland due to felling in advance of development schemes with no corresponding replacement and, recently, there have been suggestions that forestry land could be lost to future onshore wind turbine sites. Nevertheless, Forestry Commission Scotland is committed to increasing tree cover to 25% by 2075 and, in December 2010, the UK government made a commitment to a national tree planting campaign by launching *The Big Tree Plant*. This is 'a campaign to encourage people and communities to plant more trees in England's towns, cities and neighbourhoods…a partnership bringing together national tree-planting organizations and local groups working with Defra and the Forestry Commission to plant trees throughout England' (see http://thebigtreeplant.direct.gov.uk/index.html).

The Scottish Parliament and the Welsh Assembly have also committed themselves to increasing woodland cover and there has been encouraging local commitment over many years. For example, in the Scottish Borders, around 25 community woodlands have been formed since 1996 under the umbrella of the Borders Forest Trust with funding from Forestry Commission Scotland, from business, from the Voluntary Action Fund and the Scottish Government and European Community Scottish Borders Leader 2007-2013 Programme. The community woodlands are situated between Peebles in the west and Kelso in the east, with two others near Eyemouth.

In Coldstream and the surrounding area, existing woodland

has been well managed and felling programmes have included replanting, often with native species but also with mixed conifers. Recent housing developments have reduced the amount of agricultural land but not woodland and, since 2005, Scottish Borders Council's *Woodland Strategy* and its Planning Guidance Notes, *Trees and Development* and *Landscape and Development*, should ensure not just the safeguarding of trees but the planting of additional trees for screening and to create landscaped areas. People have become more aware of the benefits of trees and have taken more interest in planting and in conservation, both as individuals and as part of community groups. One example of this in Coldstream can be seen at the front and side of the primary school with the recent planting of the organic orchard and the horseshoe shaped tree area to form an open-air classroom.

What has particularly struck me during my inspections of Coldstream trees is their healthy growth; many of the older trees have dead wood in them but this is inevitable due to high winds, storm damage and old age. Soil conditions must suit them and, being on the east side of Scotland, sunshine hours are higher than the west and rainfall is sufficient without being excessive. Moreover, it is a rural area and the air is pure, as witnessed by the vigorous growth of the colourful lichen that grows on many branches on the side away from the prevailing winds, a sure sign of an unpolluted environment.

Appendix: Native Trees

1. British Isles

Alder, Common (*Alnus glutinosa*)

Apple, Crab (*Malus sylvestris*)

Ash, Common (*Fraxinus excelsior*)

Aspen (*Populus tremula*)

Beech, Common (*Fagus sylvatica*)

Birch
  Downy (*Betula pubescens*)
  Silver (*Betula pendula*)

Box (*Buxus sempervirens*)

Cherry
  Bird (*Prunus padus*)
  Wild (*Prunus avium*)

Dogwood (*Cornus sanguinea*)

Elder (*Sambucus nigra*)

Elm
  English, (*Ulmus minor*)
  Wych (*Ulmus glabra*)

Hawthorn
  Common (*Crataegus monogyna*)
  Midland (*Crataegus laevigata*)

Hazel (*Corylus avellana*)

Holly (*Ilex aquifolium*)

Hornbeam, Common (*Carpinus betulus*)

Juniper, Common (*Juniperus communis*)

Lime
  Common (*Tilia x europaea*)

2. Scotland

Alder, Common

Apple, Crab

Ash, Common

Aspen

Uncertain

Birch, Downy
Birch, Silver

Cherry, Bird
Cherry, Wild (gean)

Elder

Elm, Wych

Hawthorn, Common

Hazel

Holly

Juniper, Common

Lime (cont.)
  Large-leaved (*Tilia platyphylos*)
  Small-leaved (*Tilia cordata*)

Maple, Field (*Acer campestre*)

Oak
  Pedunculate, or English (*Quercus robur*)   Oak, Pedunculate
  Sessile (*Quercus petraea*)   Oak, Sessile

                                                   Pine, Scots
                                                   (*Pinus sylvestris*)

Pear, Wild (*Pyrus pyraster*)

Poplar, Wild black (*Populus nigra*)

Rowan (*Sorbus aucuparia*)   Rowan

Whitebeam, Common (*Sorbus aria*)

Wild Service (*Sorbus torminalis*)

Willow
  Bay (*Salix pentandra*)
  Crack (*Salix fragilis*)   Willow, Crack
  Grey (*Salix cinerea*)   Willow, Grey
  Goat (*Salix caprea*)   Willow, Goat
  White (*Salix alba*)   Willow, White
  Osier (*Salix viminalis*)

Yew (*Taxus baccata*)   Yew

Note 1: Although trees may be non-native to Scotland, it does not mean that trees that are native to England and Wales are not grown in Scotland. They are—and in great numbers. The same applies to many non-native trees introduced from abroad—they are very happy here if the climate and soil conditions suit them.

Note 2: Throughout the book, the common name for a tree is followed by its scientific or Latin name. Trees are organized by genus, then species, e.g. Black poplar, genus *Populus*, species *nigra*. It is possible to go further by referring to sub-species, e.g. *Populus nigra* ssp *betulifolia* or to varieties of species, e.g. *Populus nigra* var. *Italica*.

Bibliography

Books (Sources & Recommended Reading)

1. Aas, G & Reidmiller, A, transl. Walters, Martin, *Trees*, Harper Collins, London, 1994. <u>A useful, pocket guide in the Collins Nature Guide series.</u>

2. Amherst, Alicia, *London Parks and Gardens*, Archibald Constable & Co. Ltd., London, 1907.

3. Gilpin, William, *Remarks on Forest Scenery and Other Woodland Views*, R. Blamire, London, 1794.

4. Griffiths, John, *Old Coldstream and Cornhill*, Stenlake Publishing, Catrine, 2007.

5. Loudon, J.C, *Arboretum et Fruticetum or The Trees and Shrubs of Britain*, printed for the author by Longman, Orme, Brown, Green and Longmans, London, 1838.

6. Mabey, Richard, *Beechcombings The Narrative of Trees*, Vintage Books, London, 2008

7. Miles, Archie, *Silva The Tree in Britain*, Ebury Press, London, 1994.

8. Mitchell, Audrey, *Historic Kelso*, Friends of Kelso Museum, 2009.

9. Rackham, O, *Trees and Woodland in the British Landscape*, Phoenix, London, 2001.

10. Rackham, O, *Ancient Woodland, its history, vegetation and uses in England*, Edward Arnold, London, 1980.

11. Rodger, Donald; Stokes, Jon; Ogilvie, James, *Heritage Trees of Scotland*, Forestry Commission Scotland in association with the Tree Council, Edinburgh & London, 2006.

12. *The New Statistical Account of Scotland,* No. VI, William Blackwood & Sons, Edinburgh and Thomas Cadell, London, 1835.

Other Documents (Sources & Further Reading)

13. '*Inquiry into the adaptation of Agriculture and Forestry to Climate Change, the EU Policy Response*', *Supplementary Memorandum:* (cont.)

13. (cont.) *EU Woodland*, House of Lords European Union Committee, Sub-Committee D (Environmental and Agriculture), 2009.

14. *National Inventory of Woodland and Trees Scotland–Borders Region Part 1–Woodlands of 2 hectares and over,* Forestry Commission, Edinburgh, 1999.

15. *Native Woodlands of Scotland*, Forestry Commission, Edinburgh, 1998.

16. *Paths around Coldstream*, Scottish Borders Council, 2007.

17. *Pasture-Woodlands in Lowland Britain*, Institute of Terrestrial Ecology, Huntingdon, 1986.

18. *Scottish Borders Woodland Strategy: New Ways for Scottish Borders Trees, Woodlands and Forests*, Scottish Borders Council, 2005.

18a. Report to Scottish Borders Council as part of *Further Developments of Scottish Borders Woodland Strategy* (2009-2011), *Moving Forward with the Woodland Strategy Project 2.1., Investigation into the "Non-Spruce" Timber Resources available in the Scottish Borders*, The Buccleuch Woodlands Enterprises Limited, Selkirk, 2010.

19. *Scottish Forestry Strategy, The*, Forestry Commission Scotland, 2006.

20. *Scotland's Trees, Woods and Forests*, Forestry Commission's National Office for Scotland, Edinburgh, 2002.

21. Seed, Louise, *Survey of Border Orchards*, Borders Forest Trust, Jedburgh, 2007.

Maps (Sources)

22. Sharp, Peter, fl. 1785-1818, Plan of Coldstream, Earl of Haddington's Property [c. 1818], EMS. S.372, National Library of Scotland.

23. Sketch Plan of St. Mary's Abbey, Coldstream, Court of Session Records, Site ID RHP 49993 1589, The National Archives of Scotland.

24. Ordnance Survey 25 inch 1st edition, Scotland, 1855-1882, surveyed 1858.

www.ingramcontent.com/pod-product-compliance
Lightning Source LLC
Chambersburg PA
CBHW022020170526
45157CB00003B/1300